Brother Nature

Brother Nature

Jim Crumley

Whittles Publishing

Published by
Whittles Publishing,
Dunbeath,
Caithness, KW6 6EY,
Scotland, UK
www.whittlespublishing.com

ISBN-10 1-904445-34-9
ISBN-13 978-1-904445-34-0

Printed and bound in Poland, EU
Produced by Polskabook

To Neil and Fiona MacArthur

Other titles by Jim Crumley include:

Scottish Landscape:
Something Out There
A High and Lonely Place
Among Islands
Among Mountains
Gulfs of Blue Air (A Highland Journey)
The Heart of Skye
The Heart of Mull
The Heart of the Cairngorms
St Kilda*
Glencoe – Monarch of Glens*
Shetland – Land of the Ocean*
West Highland Landscape*

Scottish Wildlife:
Waters of the Wild Swan
Badgers on the Highland Edge
The Company of Swans

Autobiography:
The Road and the Miles (A Homage to Dundee)

Fiction:
The Goalie
The Mountain of Light

* Collaboration with photographer Colin Baxter

Acknowledgements

The author and publishers are grateful to the following for permission to use their photographs:

Cover
Laurie Campbell – photographs of osprey and wolf

Text
Laurie Campbell – pages 22 and 144
Bridget MacCaskill – pages 10, 68, 96 and 106 (photographs by Don MacCaskill)

All remaining photographs are the work of the author.

Contents

Part One
The Brotherhood

1
The Bear on the Shore

OVER THERE on the far shore, between the forest edge and the water, something moved, a creature. I saw it emerge from the forest the way an otter emerges from a Hebridean sea. That was my initial thought, as though I was untrusting of my eyes until I had first imbued the distant unfamiliar movement with the sense of something familiar.

The creature paused to drink. He raised his head then and looked towards me. I tried to catch his attention, but quietly, discreetly, so that he was not afraid. It did not occur to me to be afraid. When this meeting was foreshadowed long ago, in a childhood dream, I was not afraid. I grew up nurtured by many animal dreams to be unafraid.

This was the first time any of those animal dreams had revisited me in the real, the living, wakeful world. I had been watching the creature for some minutes before the memory of the dream returned, a dream undisturbed since I dreamed it fifty years before. Where had it rested all that time? It reached me as if from a great distance, as a voice that rouses you from a deep sleep.

But that, over there on the far shore, between the forest edge and the water, that is no dream bear. That is a young adult male grizzly bear and he looks up from his drinking and sees me and neither of us knows fear. It may be the distance of open water between us, it may be that he knows I have dreamed of him before and that the dreaming has taught me only to admire him, to be calm in his presence and not to fear.

I believe in these things. I believe that my own species and the bear's, for example, can still reach each other, still exchange basic ideas, still communicate. I believe that my species has neither wholly lost nor wholly abandoned its animal traits, that there remains deep within me a kernel as wild at heart as that bear. The difference between us is a matter of degree. I am wild in the heart of an embedded kernel, the bear in every fibre of mind and body. I feel a duty to shrink the distance between us.

I will tell him now, silently, of the old dream and ask him to believe me.

His response – if response is what it is – is to stand on his hind legs so that his huge bear head is suddenly seven or eight feet up in the air. It is an impressive stance, but he does not stand to be impressive, nor for that matter to look fierce. He stands because he is curious. He wants more information, a better sight, a clearer scent.

I send him what information I can: that my purpose is benevolent, that I am well disposed towards him, that I would like to learn from him by watching him. And I raise a hand to him and hold it aloft. Something moves between us like a warm wind. Do you feel it, bear?

He drops down on all fours again, sways his head a couple of times to left and right, then half turns and begins to walk the shore. He walks with a quiet, measured, unhurried ease, confident in his place on the map of his world. He walks without pause and without looking inland or out across the water. He walks the shore eastwards. He walks and walks and walks, a mobile fragment of the landscape, a mile without a pause. There are many more miles of shoreline ahead of him. I ask him aloud:

"What are you doing? Where are you going?"

A quiet voice at my shoulder answers:

"He is looking for his brother."

Three days ago, a small float plane landed on a lake on Kodiak Island, Alaska, and taxied towards a small island. On the tiny dock, a tall, broad-shouldered man was saying goodbye to a party of guests as they boarded a second plane. All of the guests were impeccably and expensively garbed in yellow, red and purple jackets. Their host waved them off, and turned to me as I climbed out of the first plane. He eyed my head-to-toe mud-spattered dirty green jacket and trousers (a couple of days earlier I had been knee-deep in a Yukon bog looking for trumpeter swans), and grinned a greeting:

"Ah, good," he said. "Dirty gear."

"I've been looking for trumpeter swans," I said, as if that should explain everything.

Apparently it did.

I was in Alaska to make two radio programmes for the BBC about the relationship between people and wildness, and I was on Kodiak in particular to see grizzly bears at close quarters, and to meet Scott Shelton. He had studied them for eight years. He knew them as individuals, talked to them, and insinuated himself into their presence so that they considered him a part of his landscape, a privilege they seemed content to extend to anyone who was with him, as long as they obeyed his single rule – the bears have right of way at all times. He spoke softly and he laughed easily. He was a professional bear guide, taking small groups of people at a time to watch Kodiak's famed grizzlies. He lived there with an assistant, a small, chatty and pretty woman, and they spent the spring, summer and autumn on the island. Winters, when the bears hibernated, they migrated briefly, to work with wolves in Montana.

For three days I followed while he led the way along trails through head-high fireweed, often to a stance above a riverbank. And there were the bears, fishing, playing with cubs, socialising and squabbling with each other. Occasionally one would pay us particularly close attention and he spoke to it quietly. His reassurance was all the bear needed. In the evenings, over dinner, he talked about the bears, about Kodiak, about Alaska, about the native Americans who taught him to talk to his "Brother Bear", about co-existence with wildness and wilderness.

There were three defining moments on Kodiak. The first was when I stepped too far away from Scott to take a photograph of the bears on the river. One of the bears suddenly charged at another – some domestic dispute over a salmon – and drove it out of the river and up the bear-sized footpath where I happened to be standing. I heard Scott's voice raised (the only time it happened), an urgent edge to it. But he was not talking to me. He was addressing the bounding uphill bear, and at the sound of him it stopped, and looked round. The aggressor bear was back in the water and the bounding bear turned and walked placidly back down to the river's edge.

She had come to within ten yards of where I still stood. I had not felt fear. We discussed it later.

"What might have happened if the bear had not stopped?" I asked Scott.

"You were in her way," he said. "She might have swatted you aside."

In the process she might just have removed my head from its neck. Scott was looking at his hands clasped on the table in front of him as he spoke. Then he looked up at me directly and smiled and said:

"You handled it real good."

Something locked into place then, something deep and indefinable at the time, but recognisable to me now as a kind of spiritual homecoming. I kept thinking...*I felt no fear.*

The second defining moment was also at the riverbank. We watched a grizzly sow leave the river and walk up the same path where I might have been swatted. She walked round the edge of the clearing well away from us, but in a semi-circle so that the distance from us never increased.

"What's she doing?"

"Having a look."

"At us?"

"Probably at you. She knows what I look like." (This with a soft chuckle.)

She left the arc of her semi-circle and turned down one of its radii, straight towards us. Okay, straight towards me. She stopped about twenty yards away. She raised her head above the tops of the fireweed and sniffed and stared. I filled my mind with the sight and the sense of her, Brother Bear at twenty paces. Okay, Sister Bear.

We watched each other. I don't know what she thought. I revered her, her wildness, her nearness. Remember what this looks and feels like, I told myself. It is not likely to happen again.

A few days before, up in the Yukon, near the Canadian border, I had seen my first bear from the window of a motor caravan. The driver, a Canadian biologist called Dave Mossop, stopped the vehicle. The bear was berry picking two hundred yards away. Then it vanished, moving through head-high vegetation, fireweed, birch scrub.

How can something that size vanish?

Why is there no sound?

Where is it?

Suddenly it crashed through the edge of the vegetation onto the road, bounded across and up the far bank, and with a single glance back over its shoulder at us it wandered off into the forest. Dave wanted to look at the bank for footprints. They were very clear and very large. Close by, so were the footprints of moose and wolf. Welcome to the wilderness.

But that bear had come and gone in a few moments. This one on Kodiak Island had come and was still there, still looking. And still, I felt no fear. I sent what I could out into the air towards her, my reassurance, my guarantee of good conduct in her territory.

Eventually I remembered the camera in my hand. Slowly I raised it and took a single photograph. It shows very clearly the tops of the fireweed with a bear's head parting the fronds. She satisfied her curiosity, turned, followed her own path back to the semi-circle, and returned to the river.

"That," said Scott, "should be a nice picture."

But I never saw him take one himself. I guessed he was a walking filing cabinet of bear images, and had no need of photographs.

I asked him later how close he'd ever been to a grizzly. He told me this story:

"I was walking back through the fireweed to my boat, and was almost at the shore when I heard noises, as if someone was going through my boat. I reached the edge of the fireweed where the boat was hauled up and saw that it was a bear that was going through the boat, looking through my gear, searching for food. We surprised each other. Our heads met.

"Her ears went back, and I thought she was coming for me. But then they popped forward again and I knew it was alright. She spared me."

"How close?" I asked.

"Oh, about a metre."

The third defining moment in Alaska, then, was that last evening when the bear emerged from the wood, looked at me across the water, and began to walk and something moved between us like a warm wind.

Do you feel it, bear?

And as he walks the shore, I ask him aloud:

"What are you doing? Where are you going?"

A quiet voice at my shoulder answers for the bear:

"He is looking for his brother."

I turn to the voice, Scott's voice, and ask:

"His brother being...?"

Scott shrugs, opens his arms, gestures with them at the rest of Alaska, turns and walks back to his cabin. I turn back to the bear. He has walked perhaps a mile since I first saw him. He still walks. Neither the pace nor the direction falters. He is not hunting, he is not exploring, not pausing to check scents. He is on the move, and the open shoreline is an easier highway to travel than the forest. So he walks the shore. He walks and walks, and I watch until he is too small to distinguish from the shoreline boulders. It is all a brotherhood. Bear, shore, forest, water, mountain, salmon, wolf, caribou, moose, kingfisher, loon, lynx, mountain lion, eagle, fox…whatever. And here and there, in the wild world, there is a man or a woman who has re-learned and reclaimed the right to our own species' old place in the brotherhood.

So I rest my eyes after the effort of watching a bear walk a few miles of shoreline, and something occurs to me. I have watched him just being, just going about his business, he registered my presence only fleetingly, and probably because of the distance of open water – just possibly because he felt the warm wind – he has accorded me the status of a fragment of his landscape. And equally possibly, because he has associated me with Scott's place, I have briefly borrowed a tenancy of the brotherhood. And that, it occurs to me, is what Scotland was like a thousand years ago. And that is at least a part of the great worth of the Alaskas of this world. They matter firstly, of course, for their own precious and primitive sake. But they also have the capacity to teach those of us who wash up briefly on their shores that if people are willing to make the effort, conserve, restore, co-exist, then a closer walk with nature is not beyond us, and we have nothing to fear from the consequences.

A few months later, just as my radio programmes were about to be broadcast, I heard that Scott Shelton had suffered a stroke and died. I believe utterly that when the bears emerged from hibernation the following spring they would mourn the absence of their Brother. Scott was a bridge between what we have become as a species and what we could have been had we not chosen to set our face against nature, a bridge by which understanding might cross. When I heard of his death I felt sorrow for the bears' loss.

2
The Badger on the Shore

THE BADGER IS ALL the bear Scotland has known these last one thousand years. But the Caledonian brown bear was known to the Romans for its fearsome reputation in a tight corner. They captured hundreds and shipped them back to Rome from Berwick, a name cobbled together from "bear" and the Norse "vik" – a haven, a bay, a port. The bears put up a great fight at the Coliseum, especially the females. Today, the loathsome badger baiters know there is no tougher test of their terriers' mettle than a badger, especially a female. It puts up a great fight in a tight corner, a crude coliseum of the badger baiters' making.

Badgers have the bear's inswinging gait, the bear's casually packed power, the bear's supersensitive snout and indifferent sight, the bear's prodigious digging machinery, the bear's omnivorous versatility from berries and worms and wasps to poultry, carrion and fish.

Fish?

I quote from the badger-watcher's bible, *The Natural History of Badgers* by Ernest Neal. Note the two words that precede the word "sweep" in his account of a badger-watcher's homeward journey by a river at one in the morning:

"...he suddenly saw a fish appear above the grass and a moment later heard it fall and flap about on dry land.

"On approaching the river silently he was astonished to see a fully-grown badger standing on a solid mud spit which projected nearly two metres into the water. It had its right fore paw raised and was staring intently into the water at its feet. Suddenly it made a bear-like sweep with its paw and neatly flipped a small fish out of the water and onto the bank behind it...Then the badger...turned and ate the fishes on the bank and wandered off."

I remembered the bears in the river, the sows fishing for salmon that flowed downstream in red tides, so many salmon that they were ripping them open, eating the eggs then tossing them to the cubs. The cubs ate and left the remains on the banks. Magpies queued up for the spoils, and merlins raked the treefuls of magpies in lethal low-slung forays. I know a Scottish gamekeeper who hates magpies. I told him Kodiak Island is the place to go to see them die.

I remembered the bear on the shore that stepped from a childhood dream, and I looked up to see a badger on a loch shore of the Highland Edge. I saw it emerge from the shoreline trees the way an otter emerges from a Hebridean sea. It crossed the stone shore to the water's edge and drank briefly, pawed at something in the water that proved to be the long dead remnant of a pike, thrust his nose at it, then leaped backwards in disgust with all four feet off the ground to land with an undignified thump among stones and tree roots. Then it raised its head, and waved it from side to side, sniffing the air, and it was a pocket-sized bear. It turned, bear-like, and walked away along the shore, looking for its brother.

And if you were to ask me what I mean by that I would shrug, open my arms and gesture with them at the rest of Scotland's Highland Edge. For it is all a brotherhood. Badger, shore, forest, water, field, foothill, mountain, pike, trout, salmon, deer, wildcat, fox, pine marten, otter, red squirrel, red kite, osprey, golden eagle, whatever. But in this of all landscapes I have no need to borrow the tenancy of the brotherhood from another, for this wedge of the terrain I call the Highland Edge has been my working territory for thirty years. I have long since earned the right to my tenancy of *this* brotherhood.

The words "The Highland Edge" appear on no map, and that territory I love to work conforms to no known geometrical shape. It is a ragged slice of mostly Perthshire from the low-lying Lake of Menteith in the south to the mountains of Balquhidder in the north – a little less than 20 miles – and perhaps a dozen miles from east to west. But like any creature's territory in such a landscape it is a flexible realm, varying by a mile or two with weather and season and whim. But the core is constant, and that would have to be the country around Loch Lubnaig and Loch Voil. One lies north-south, the other east west, and the north end of one and the east end of the other are linked by the River Balvaig. Their surrounding hills and forests and mountains fill the window where I write. This book is written from within. Loch Lubnaig is boomerang-shaped and its northern end is split into two bays by the last long thrust of the River Balvaig's tree-lined banks. There a pair of mute swans nests in circumstances as trying as they are untypical. I have followed their generations and their struggles, their copious supply of natural disasters, their few glittering triumphs. And from these birds I have learned how to be still; learned something of the heroic nature of wildness, of the stoic nature of the brotherhood, of the chains that bind all the brotherhood from lowland loch shore to mountain summit.

And I have guessed thoughtfully about absences – pinewood, lynx, boar, beaver, Brother Bear of course, and wolf. As I write this, the bureaucratic talking shop that passes for Scotland's conservation decision makers has been discussing the reintroduction of the beaver for about a dozen years. The beaver remains an absence although sites for a controlled reintroduction have been identified here and there across the Highlands, then discarded. The bureaucrats haver on and landowners and fishermen bleat wearily about perceived threats that do not exist and never have.

Meanwhile, a landowner with 23,000 acres to play with in the northern Highlands wants to throw an electric fence round it all and introduce the wolf and the lynx and the boar there, which is to misunderstand and misrepresent hopelessly the nature of wildness, and especially the nature of the wolf. But it is the kind of well-intentioned misguided substitute for wildness that happens when the core arguments get suffocated by bureaucrats.

About the beavers: ten years ago I made two radio programmes for the BBC about our extinct wild mammals and how they might be reintroduced. One of the people I interviewed was Don MacCaskill, a retired forester, a wildlife photographer of exceptional skills, and an instinctive and uncompromising naturalist. For the latter part of his working life in the Forestry Commission he was the chief forester at Strathyre Forest, which includes the hills around Loch Lubnaig and Loch Voil. So his personal imprint was a legible element on the landscape of my Highland Edge territory. And he had a theory about beavers.

So we stood on the wetland shore of Lochan Buidhe, a reedy little watersheet to the north of Loch Lubnaig frequented by otters and herons and wintering whooper swans, and an occasional resort of the mute swans from the loch. There were four of us that spring morning, Don and I, BBC producer Dave Batchelor, and a man who worked with beavers in France, and I interviewed my old friend. It was he who had brought us to the shore of Lochan Buidhe.

Here, he said, was where we should introduce beavers.

Why here? I asked. I expected a naturalist's assessment of habitat, food supply, lack of disturbance, that kind of thing. And perhaps a photographer's assessment too, for he lived nearby and he would have loved to be involved in such a project. But instead he said:

"I had a dream they were here."

At which point the man from France smiled and nodded vigorously and said:

"I am impressed. Do you know, this is *so* like the landscape in France where we reintroduced beavers?" He nodded again. "*So* like it."

I go often to Lochan Buidhe and almost as often I think about Don's simple conviction that his dream validated the idea of beavers here. And the more I think about it the more it impresses me too. Because there would have been a time – even among these very hills – when the assistance of dream or second sight or some knowledge acquired intuitively from some realm beyond reason would have informed decisions and determined the course of events. Seers and sages, sung and unsung, are part of the history of such places.

North American Indian and Inuit cultures are rich in such traditions, and we often know more about them because they stayed closer to nature for much longer than we did (and in some places they still do), and because America's admirable tradition of nature writing has written them down and found a market place in modern popular culture. Don lived his life in nature's company. He grew up at Kilmartin in Argyll where the awareness at least of such traditions is perhaps better preserved than in most other parts of Scotland. And if you spend your life working with nature on nature's terms, determined to live as close to it as is humanly possible today, it is far from fanciful to suppose that nature in return might trust you and use you to further its cause, and insinuate its purpose into a dream.

And if you take that philosophy and present it to such a body as, say

Scottish Natural Heritage, or the petitions committee of the Scottish Parliament, and suggest that it is a more appropriate basis for action on behalf of wildlife than their ponderous bureaucracies, how far do you think you would get?

And yet consider this. The dream was simple and specific – a particular species in a particular landscape. The pedigree of the dreamer was impeccable. His knowledge of nature was gathered in nature's company, much of it in the landscape of his dream. On the basis of that knowledge, he was permitted to design forestry plantations and change the face and the ecology of mountainsides. He was accustomed to thinking deeply about reintroducing species, especially those that inhabit a treed environment. I had heard him discuss the notion many times. If that substantial store of intimate knowledge and careful reason was then infiltrated by a simple and vivid dream, why would you not be justified in trusting it utterly as a thing of profound significance and defining purpose?

I dare them.

I dare the ponderous ones at SNH and the occasionally surprising individuals at the Scottish Parliament to act on Don MacCaskill's dream. Let them have their scientific trial by all means, radio collar and ear tag their beavers if they must, and follow their every waking, sleeping moment and cull them if they get out of hand or if the project takes cold feet, make their scientific appraisals and chunky reports and collect their salaries. But allow the dream too. Allow beavers to be wild here on the Highland Edge, untagged, uncollared, and allow nature to make the decisions, and leave those of us who still believe in the natural order to watch and wonder at what unfolds, at the revealed purpose of nature in choosing a man like Don MacCaskill to be the seedbed of a dream.

Meanwhile, ten miles to the south, beyond the first and last of the mountains, beyond their foothills, I had watched a badger on the shore of the Lake of Menteith scent the air of an early May evening and go looking, bear-like, for his brother. The badger is my good omen in my adopted territory. When I first came here from the east in the early 1970s, the old setts and badger highways were cold. Many hands had been turned against them, setts had been gassed by the kind of old school gamekeepers who used to nail a pair of hen harrier wings to a barn door as proud evidence of their good stewardship of the land. One conifer woodland sett in the foothills had been gassed by the Forestry Commission although no-one could ever explain to me why a forester would find the practice beneficial to the forest. It would not have happened in a forest managed by Don MacCaskill, but then people like Don often found themselves at odds with the system in which they were required to work. Perversely enough, the gassed forestry sett was in a clearing and the decision was taken not to plant the clearing in case the badgers ever returned. There are many thoughtful and enlightened people working for the Forestry Commission on the Highland Edge, many more working for Scottish Natural Heritage, but often their idealism is stifled by a kind of institutionalised madness.

For at least twenty years, perhaps longer, no badger stirred. I didn't miss them because I had never known them here, and in this place of overlapping Highland and Lowland landscapes nature appeared to be in good heart, and its tribes were everywhere on the march in woodland and on water, in foothills and mountains, and in all their skies. But slowly, over the years, I began to stumble across the badgers' lost civilisation, in the way that an archaeologist might piece together the lives of people banished by the Highland Clearances from the West and North, putting flesh on the ruinous old stone bones of the homes they left behind, the artefacts of their civilisation that lay buried in the peat, the songs, the poems and the stories they handed down on both sides of the Atlantic. And for a while it seemed that every time I explored a south-facing slope or woodland edge with a burn or a ditch nearby I would find evidence of old badger haunts from whole setts to solitary outlying holes, and the clawed signatures of the badger tribe on standing and fallen trees. It was not just that there had been badgers here. There had been *a lot* of badgers here, and the campaign to eradicate them had been determined, widespread and presumably co-ordinated. The answer to the question why remains a mystery.

I considered the situation in a book I wrote in 1995 called *Badgers on the Highland Edge,* and suggested that the position could be reversed by lifting a number of badgers across the central belt of Scotland and resettling them in the old setts of the Highland Edge. If that is what happened, it was done with stealth rather than official blessing, but I don't think that was what happened. Because even as I was writing that book, something had begun to give, and the grapevine that attends the endeavours of those who labour in nature's cause began to cough up the first symptoms of a change – a glimpse of a solitary badger in headlights, a road kill; and then one day when I was looking for something else altogether, I stumbled across what is still the biggest badger hole I have ever seen, and with a newly excavated spoil heap of dark earth. I must have passed it a hundred times during its unexcavated years without knowing it was there, cold and overgrown and forgotten. It was an outpost, probably a temporary refuge for a solitary wanderer, but it suggested that at the very least badgers had begun to dip a toe in once familiar waters. But it had been twenty or thirty years, and no badger lives that long in the wild. So what persuaded the tribe to think again about the Highland Edge? Race memory? Had some store of accumulated wisdom been handed down through several badger generations? Had something fundamental changed here, something of crucial importance to badgers but unobserved by the badger-watcher? They couldn't know the old keeper was dead, or that they were likelier to receive a more considered welcome up in the forestry sett. But for whatever reason there was a tentative presence again where there had been a long and enforced absence.

I was discussing the situation with a veteran of many badger campaigns. He considered my theories carefully and in silence for a while, then he shrugged and said simply:

"Or...once a badger sett, always a badger sett."

The clearing lies on the southmost face of the Highland Edge. It slopes gently, and in high summer it is impassable with head-high bracken. It is surrounded on three sides by huge trees, beech, oak, larch, Douglas fir, sycamore, birch, ash, aspen. But to the south it is open, and borders on marginal farmland, beyond which the world lowers and widens into the green carselands of the Forth. The sett is in the crown of the clearing. This was the sett the old keeper gassed. When I found it first I had to claw away armfuls of dead bracken to find evidence that badgers had ever been here. But some of their old paths had been kept open by roe and red deer, foxes, occasionally otters on the march. I had found the first two holes by following the paths. I discovered half a dozen entrances within about twenty square yards, and here and there beneath the eager bracken were massive earthworks. The badgers had been here long enough to reshape the contours of the clearing. It was a long-lived civilisation that had been terminated by cyanide.

But look at it now on an evening of early May. Behold the transformation. The main sett has twelve holes in regular use. The paths that link them are flattened and laid bare by the passage of badger feet. New earthworks sprawl downhill and at the core of the sett around those original half dozen holes that grew cold, the new spoil heaps are wide and steep and split by a single conspicuous groove. That groove is suggestive of the boisterous cut and thrust of cubs at play. Two hundred yards away to the west a new outstation was opened up a few years ago, and now new holes have begun to appear there too. It is the same to the east of the clearing where an old hole beneath a single birch tree has been re-opened. And everywhere around the clearing are badger latrines, areas liberally dug and dolloped with dung pits. If you want to know whether a sett is busy or not, count the number of dung pits.

But I don't count. Translating wildness into statistics was never my idea of a close relationship with nature. I don't know how many dung pits there are, although I know there are at least three latrines in different parts of the clearing. I don't even know how many badgers use the sett, and I don't care that much. I know that it thrived long before I came here, that it was made cold and the badger tribe stayed away for twenty or thirty years, that it thrives again, produces cubs and continues to expand. And now I realise how wrong I was in those early years to settle for a Highland Edge without badgers. Because now that they are back, now that I have watched them re-stake their old claim, recolonise their old heartlands, reoccupy their old place in the landscape's scheme of things – and all of that unaided – I realise the impoverished nature of the wood without them.

And not just the wood. It seems they abandoned their remoter niches on the Highland Edge about the same time as the gassings, which in turn suggests that the big setts supplied the badgers for the remoter ones in the high hills. But now they are also back in the forestry sett in the foothills, and back on the mountainsides of Balquhidder both in a mountain woodland around a rockfall and also out on the barest of hillsides at 1500 feet. Piece by piece I have assembled my idea of how they move around the Highland Edge, but why some will choose the shelter of the southern clearing and others the storm-strewn bare mountain slopes of the north is beyond me. But in the same way, the red deer is at home both in the southern wood and in the snows of mountain watersheds.

So, it is an evening of early May, and I have been sitting just below the crest of one of the clearing's many small mounds, the wind in my face, and the core of the sett fifty yards away. At this time of year, the badgers are out and about in the long daylight of evening. The clearing shadows early because of the encircling trees, but the light is in the sky for hours yet and sunlight still shows on the hills to the south of the carse.

A big boar badger sniffs the evening air from just inside his usual exit hole. He is the roost-ruler, the he-who-must-be-obeyed of sett society. He emerges into the lit overworld and stands. He looks around and scents every air at his disposal. He scents and hears far more than he sees. He looks at my gray-green shape which he knows is sometimes there on the mound and sometimes not, but it doesn't look like what it really is and the wind bears its scent away from him. He looks hard and my binoculars look back. He looks away, then moves off at an easy trot, heading for the woods.

A roebuck barks somewhere behind me, a constant element in the vocabulary of the sett along with the squeak-and-grunt of woodcock, the plaint of buzzard, the hours-long silvery flourish of mistle thrush, the havering of cuckoo.

There are more badgers now, four of them scenting and nuzzling and muttering close together above the main entrances. One by one they choose a path and embark on the night's business. Another appears in the furthest corner of the sett, and from where I sit there is no way of knowing if it has used a hole on the edge of the main sett or has wandered into the clearing from the outstation to the west. I had been hoping to see cubs, but they were elsewhere for the duration of the vigil. Some nights you see them over and over again, oblivious and uninhibited, their play and their explorations all but spilling over your boots where you sit, which is your reward for the stoic qualities of your sitting. During others, not a single badger shows, or they stomp off into the woods and are gone for all the hours you can thole, and that is part of the price you pay. But you rarely leave the sett completely empty handed.

The roebuck barks again. He is closer now and directly behind me. Suddenly it occurs to me to pay him some attention. While I had been preoccupied with the badgers his voice had been an occasional offstage voice. Suddenly it is onstage and closing.

Consider this. My seat was carefully chosen – a slope on a mound, far enough down the mound so that I wouldn't present a silhouette against the sky to the badgers, and downwind from the badgers so that I wouldn't present a troubling scent. All of which was now troubling the roebuck, for he was downwind of me, and troubled by the fact that he could smell me but not see me, for he was on the wrong side of the mound. The last thing I want now, either for my own comfort or the roebuck's, is for him to walk over the top of the mound in a state of some agitation and find me sitting there a few feet below him. On the other hand, I'm reluctant to move because I've done the hard part of the evening's work, been still enough for long enough to have become a fragment of the badgers' landscape.

The roebuck barks again. Louder, more aggressive (it seems to my ears), and definitely closer. I imagine him looking over his shoulder to the woodland where perhaps there is a doe and a very young calf; then he turns, advances a few paces more towards the source of his disquiet – me, though he doesn't know it yet – stands and barks his warning.

The next bark is just too close for comfort. Time to let him see me. I'll talk to him, Brother Deer, hopefully reassure him. More than likely he will retreat. More than likely…

I turn in to the face of the mound and very slowly, I stand. The sight that greets my eyes as my head clears the top of the mound is one that will endure in my mind for as long as my mind endures.

He stands not twenty feet away. His summer red-brown coat glows in the dusk. His six-pointer antlers catch some stray splinter of light that renders them dusky gold. And his legs are planted thigh deep in a gently undulating sea of wild hyacinths that drowns the eastern end of the clearing and swirls away in among the trees as far as my eyes can travel in the half light. And there he stands. For perhaps ten seconds he stands, and no fragment of him moves.

I wonder what he feels. I feel sublimely supercharged. Some force pulses up my spine and throbs across my shoulder blades. Very slowly I raise my right hand until it is more or less level with my shoulder, palm towards him.

Why would I do that?

What could it possibly mean to him?

I want it to be a gesture of reassurance. I tell him aloud that he has nothing to fear from me. That I admire him. I want him to understand what I mean by the gesture. I believe he is capable of understanding that. Brother Roebuck.

He barks again. Its suddenness and its volume go through me. I think he tells me that he understands the gesture, but that he fears me anyway.

He turns sideways on. A ripple of the blue sea courses about his legs.

And he stands again, looking at me, then back across to the wood. He barks at the wood. The unseen doe answers, a softer, gruffer bark. Then he turns and runs for the trees, splashes among the wild hyacinths, stirring their hyacinth smell as he goes. But just before the edge, he stops, turns, stares, barks twice, turns again and gives himself to the woody sanctuary.

There is nothing left of the evening. I gather up the mat I was sitting on, a flask, a notebook and pencil, put them in the rucksack and walk softly out of the clearing avoiding the sett, avoiding the badger paths.

My car is near the shore of the loch. There is still light there in the western sky and I linger for a few minutes looking out over the water. My head dreams of the deer up to his thighs in the blue sea of flowers. Then I remember the old dream with the bear in it. A soft noise interrupts me. I look up to see a badger on the loch shore. I see it emerge from the shoreline trees the way an otter emerges from a Hebridean sea.

The badger is back on the Highland Edge because the tribe judged the time was right. Once a badger sett, always a badger sett. The land is immensely enriched because he returned. The beaver cannot make that kind of judgment because there is no native population. We alone – your species and mine – can permit the beaver to return and be wild here again. There is no reason on earth to deny it a day longer. The land will only enrich by its return.

3
Nature's Sacred Hour

THE LOCH LIES as level and still and gray as ice. A young mute swan, gray-brown and rumpled and a little less than a year old, scoops busily at something just below the surface as it slowly swims. The tiny ripples of its wake subside in a few yards and the ice-like stillness closes in again behind it.

A raven calls far across the loch. In the silence that follows its echo bounces and reverberates like a stone thrown across ice. Even the colour of the loch suggests ice, blue-gray with hints of suppressed light. But it is late March and far too mild for ice. The raven picks up where it left off and begins to call and call, relentless as cuckoos. A great-crested grebe, newly arrived from the winter firth but clearly dressed for the unborn spring, stares at the source of the sound then begins to call back, a harsher, lower-pitched guttural. Does it really think it hears another grebe, or does it know the raven for a player of games and is it playing games too? Or is it coincidence?

The voices of distant pink-footed geese rummage between the sky and the unseen farm over my shoulder. Two adult mute swans doze through the afternoon stillness a few yards from my shore, a companionable pair of curves. Occasionally an eye blinks open and watches me from the middle of a swan spine, which is where the bird happens to have stowed its head. Mist and low dark clouds hug the hills and hide the mountain. The loch is at ease. The year waits.

I have succumbed to the mood of the hour. There is a writing pad on my knee and a pen in my hand but the pen has been still for some time and the last sentence lies half-finished, its purpose unrealised. Then the branches of an overhanging Scots pine shiver. My back is to the pine trunk and something transmits itself through my spine.

What just passed by?

The loch shallows sizzle softly under a shower of raindrops the tree had harvested in the windless hours of mist and drizzle. The breeze that caused the fall and stirred the pine now causes the loch to whisper, then to mutter, then to gossip aloud, then to dance.

Every tree sighs and stirs.

The dozing swans rouse.

The sky leaks sudden sunlight.

The water brightens to something like tinfoil as it dances, begins to slap the shoreline rocks.

The swans converse and turn and swim.

A horde of geese – perhaps two thousand birds – crosses the trees and falls in a loud and frantic cloudburst on the far shore.

The swimming swans startle as a third swan emerges from a nearby bay, flying. The sound of its wings batters back into their ears and mine. The flier crash-lands noisily a quarter of a mile away.

The wind gathers and begins to shout, tears holes in the clouds, rips them from the hills.

The mountain stretches like a waking wolf, bares its long, blue shoulder.

I stir myself, gather my thoughts and writing materials. I have business in the badger wood behind me. I think I have felt the first breath of the new spring.

Beyond the clearing that accommodates the badger sett, the surrounding woodland climbs to a low ridge. Native trees defend the woodland fringes and move in on the open spaces, but deeper into the wood there is the unmistakable signature of Victorians at work. The big house may be decrepit, and much of the wood stifled and darkened by rampant rhododendrons, but rising far above it all stands the supreme validation of Victorian endeavour here, a dozen monstrous trees. They are mostly Douglas firs, all of them well over a hundred feet. Victorian landowners were zealous gatherers of the seeds of foreign trees, especially big showy foreign trees; the grand proportions of the trees topped off their grand designs. These firs stand in two groups a couple of hundred yards apart. In the topmost branches of the very tallest, and all of 150 feet above the ground, there is a nest as big as a swan's, but many times heavier, for it is made not of reeds and grass but of hundreds of sticks from twigs to substantial branches. Every winter's gales hurl the whole colossal structure to the woodland floor and make matchwood of it, and every following spring the returning ospreys rebuild it and begin again. In a big wind the nest reels yards across the sky, and big winds are far from uncommon on the Highland Edge in the birds' early spring nesting season. But this particular tree has proved to be among the most fertile of all Scotland's osprey eyries for over twenty years. Its offspring have helped to establish a new dynasty of ospreys within a radius of two or three miles. Now, almost every other year there is another new nest, another new breeding pair.

The ospreys lured me here during the years of the badger-coldness. They used to nest in a dead pine, a 20-feet high runt, but the surrounding trees outgrew it and they lost their view, which is the one thing every osprey demands of its nest site. But every late March from then till now, which is thirty-something years, I have followed a narrow little footpath of my own making to the top of this bank among the whins from where I can scan the roof of the woodland. Sometimes the year's first glimpse of a returning osprey is out over the loch. Sometimes it is already sitting on the nest tree, wearied by the homecoming flight from west Africa, contemplating the ruination of one more nest. Sometimes it drifts in over the trees with a trout stowed in its talons. Sometimes it is hunched against a blizzard. And just once, just this once...

...The branch looked dead at a distance, as though it would snap off at the slightest pressure. A good branch like this one is essential for the job in hand. It looks to be about five feet long, two or three inches thick. It looks so long dead it will surely snap like a matchstick.

Take a good grip. Apply serious downward force. Anticipate the clean snap.

Nothing.

The branch holds firm. It is not just as dead as it seemed at a distance.

Try again. Strengthen the grip, increase the downward force. This time.

Nothing. The branch stays put.

Rethink the strategy. That branch is very important. Jump on it from a great height. That ought to work. It doesn't, although there is a sharp crack, the first hint that its resistance is weakening. Close examination reveals that near to where it meets the trunk, it has cracked about half an inch down and four or five inches sideways.

Take a good grip again, and bounce, holding on, and as it cracks again throw both wings high into the air and start flying but still holding on to the branch with both feet. That works. The branch gives, and with remarkable control considering what has just been achieved, the osprey shrugs off the migratory weariness and manhandles the timber down towards the ground.

Surely such a burden can only travel downwards. Surely, if the thing ever touches the ground, no bird on earth will ever get it airborne again. But I don't think the way an osprey does, nor do I know what an osprey knows about manhandling timber, nor do I know how to manipulate a canopy of wings so that its flexibility extends from long-haul flight to the demands of vertical timber haulage.

A few feet from the ground, the bird levels out, and begins a long, slow turn that evolves into a long, slow, *climbing* turn. Consider what the bird now proposes. Look up to where last year's nest was, in the very topmost branches of that 150-feet high fir, the highest tree in the wood. What the osprey has in mind is to get his timber from more or less ground level to the top of that tree.

The laborious turn becomes a full circle, wings heaving. It becomes a second circle, a third. The fourth takes the osprey to where it needs to be, which is about ten feet higher than the treetop. From there, it circles on until it can approach the tree into the wind.

It begins, at last to glide, to descend, and, improbably, to slow almost to a stall. When it reaches the treetop, it is hardly moving at all, its wings rise up and back, and it lays the dead branch amid the wind-troubled greenery of the living treetop the way you might lay a baby in a cot.

So, on this first day of the infant spring an osprey lays its new foundation stone, and begins again. It will find much of what it needs on the woodland floor, but every now and then, it goes to those extraordinary lengths to remove a carefully selected piece of timber from a dead tree. And I am quietly pleased to be a slave to this spring ritual of my own that greets the homecomer.

Perched on the top of a bank where a screen of whins only blunts the now disagreeable wind, I moved for the first time in 20 minutes. For most of that time my binoculars had been up at my eyes, my elbows resting on my knees. Those 20 minutes were the culmination of three hours of sitting on the bank among the whins, scouring the cold, mobile, moist air for something to celebrate, my first sighting this new spring of the returning ospreys. I now celebrated both the sighting and movement that permitted blood to flow again through cold, stiffened limbs.

It had been 50 years – the spring of 1954 – since ospreys were first discovered nesting again in their old Speyside stronghold of Rothiemurchus. The Victorians had more or less wiped them from the face of the land for reasons best known to themselves. (Strange creatures the Victorians – they wiped out the bird but unwittingly provided for its return with their tree-planting.) But here on the Highland Edge, the gap had been nearer 100 years than 50, and they had not found their way back until 1970. So these last 30-something years I have been accustomed to spend my March-into-April days in this wood and on that loch shore, to welcome the ospreys home. The ospreys are indifferent to my attentions. This is a need in me, not in them. Their return marks a watershed in the wild year, and the wild year is the calendar by which I order my days. Certain events that recur each wild year help to underpin my precarious writing life. They impart confidence, supply sustenance, reinforce commitment. I work with nature, rather than with other people, and nature doesn't much care whether I turn up for work or not. What matters is that nature turns up for work. So for me, the return of the osprey is a pivot on which the wild year turns again.

The osprey fares well on the face of the land again, 200 breeding pairs and rising. The loch now sustains seven nesting pairs and any number of passage birds, but their increasing familiarity does not diminish the significance of their arrival in my wild year. Rather it serves as an example. It teaches me to be willing to work at nature's pace. The naturalist David Steven once told me the single most important piece of wisdom he had ever heard. It had been imparted by a young priest, he said, and its simplicity so impressed him that he lived by it for the rest of his life. It was this:

"It takes three years to make a three-year-old."

The osprey, the eagle, the swan, the otter, the badger, the deer – all the wild tribes of this landscape – already know that, and put it into practice every living, breathing day. The beaver would arrive with the same knowledge already implanted.

On the top of my bank, the sodden wind was getting the better of the screening whins, the eyrie tree blurring as the squalls thickened and joined forces.

Fifty years, I thought, as I began to walk out to the car and the notion of a pub fire. It is only the osprey watcher that marks such an anniversary. The osprey itself marks only the turn of the wild year, one more step on the long march back from the rim of extinction that lasts forever. It is the least I can do to try to keep in step.

High-summer-blue drenched the air and the water, the day as laden with warmth as a bee bowed down with its burden of pollen. Half a mile out I cut the almost silent electric outboard and began rowing; half a mile more and I let the boat drift and began to scrutinise the sky, the trees, the surface of the loch, and all the spaces in between. It took thirty seconds to find the first osprey, quartering the shallows along the reedy south shore from perhaps a hundred feet up. I nudged the boat that way with the oars and let it drift again. The ospreys hereabouts are well accustomed to small boats for the loch is also a well-stocked trout fishery. Anguished fishermen tell bar room stories of the osprey that dived in fifty feet away and pulled out a fish it could barely lift, and they had caught nothing in half a day of trying. Fists were shaken, threats and oaths and abuse were hurled. But that blue midweek afternoon was quiet. I rowed alone and I don't fish. I watch the osprey fishing because I love to watch the osprey fishing...

...The bird quarters its chosen airspace in slow circles. The sun glitters on the white plumage of head and breast and underwings. The eyes stare down. The water is shallow, the air clear, the conditions perfect. The bird flips onto a wingtip, sets its wings in a tight "w" shape and falls. Three or four feet above the surface it pulls out and whips back up into the very airspace it has just vacated, resumes its patient circling. Ten minutes and six dives later it has still not breached the surface. With every osprey dive, the osprey watcher soars. The power and beauty of an osprey at work is what has won all our hearts and minds and made it an ambassador for wildness.

The fact that it reintroduced itself, that it has prospered again in its old haunts, that it will accept man's gifts of artificial eyrie platforms, that it returns to our skies as reliably as spring itself, that it leaves an emptiness behind when it returns to Africa at the end of summer...all these clasp the osprey ever deeper into our affections. It is as if the osprey has been building bridges towards the species that tried to wipe it out, and now that our species understands the basic principles of nature conservation, we welcome the new intimacy between bird and man. Scots may revere the golden eagle above the osprey, our mile-high clan chief of mountain wildness, but for some people the eagle is beyond their grasp, *too* aloof, *too* wild, a shunner of man. My own relationship with the golden eagle is almost idolatrous, but I understand the reservations of those who balk at the idea of it as some kind of national symbol. We must relate to the eagle on its terms or not at all. Intimacy plays no part in it. But we know now that the osprey is willing to meet us halfway, so we too are bridge-building, grateful for the osprey's forgiveness.

A second osprey cruises directly over the drifting boat, looking down from the eye of the sun. With one hand raised to shield squinting eyes I can see a bird rimmed in gold light. I have one telling moment of awareness of what a surfacing trout may go through looking up and seeing the huge hovering of pale wings. The bird flies on, and a splash resounds behind me. The hunter bird has struck. It settles briefly on the surface, heaves its wings back with their leading edges pointing at the sky, heaves them forward so that their primaries almost touch several feet in front of its face, and as they part and heave back again the bird rises and begins to travel forward. As it clears the water, its lowered legs and bunched talons haul a trout from the water. It will die somewhere over the trees, which cannot be how a trout expects to die. The second bird repeats the manoeuvre within yards of the first and succeeds at the first dive.

Far out on the loch, the surface was as still and unmoving as blue glass. Great crested grebes and cormorants sailed by on top of their reflections oblivious to the silent boat, strange birds with unintelligible languages, creatures from a dream. The sun poured down and lit their waterworld in primary colours. I watched a grebe wrestle an eel for several minutes before the eel uncoiled just long enough for the grebe to inhale it. How does *that* feel, to be sucked head-first into the darkness of a bird's innards, dying inch by inch?

There were more ospreys, in the sky, perched in trees, calling from the shore in thin reedy whistles, tipping over to smash the plate glass of the loch, climbing from the surface with and without fish and shimmying the water from the their plumage in sun-starred cascades. The whole long afternoon was an intoxication in blue and yellow and green. In three hours of rowing and drifting and watching I saw fifteen ospreys. Some birds I may well have encountered more than once, but I had six in the sky at the same time, and I rowed ashore feeling as if I had strayed into a parallel realm of nature where different laws held sway and different priorities shaped events. And when I stepped on to land again and hauled the boat up the shore and turned to look back at the loch I was already apart from it, already somehow diminished. I had borrowed from the birds, from the water, from a rarefied moment of nature, and my time was up. I had been allowed a glimpse of something extraordinary, but I had no use for it. It was quite beyond me. I could watch it but not participate in it. Even now, writing it down and passing it on almost feels like a kind of betrayal. What do I really know of what I saw? What part of my understanding of nature was enriched? I really don't know.

I drove the short distance back to the beginning of the small footpath up to the bank with the whins and the view of the eyrie tree. This was *my* path, *my* familiarity, *my* certainty. I sat where I always sit and soaked in the last of the sun before it dipped behind the trees. An osprey stood on the highest fronds of the highest tree's highest branch facing the sun, its white breast feathers glowing an outrageous shade of peachy pink. Every now and then it bowed its head to its feet where a freshly caught trout was being reduced to scraps of skin and scale and bone. All this I understood, the rituals of the nest and the tree and the wood and the home-coming and the leave-taking. But that out there, that technicolour waterworld of a hot summer afternoon, where the osprey went about its workaday business and in the process transformed into the tribe of dream-birds…that is a gap I have not yet been able to bridge.

Every new spring I put my ear to the grapevine. If you know how to tune in and if you understand its cadences, you learn. But you must also participate. You feed the grapevine too, and it feeds you in return. Information is exchanged – sightings, clues, what-ifs – and then you must confirm the information for yourself so that you have earned the right to the knowledge, or if you can show it to be false you feed that back to the grapevine too so that the others who care and participate don't waste their time. You earn the trust of the grapevine over the years, and if you fail it, it cuts you out. It is a good system.

So one more spring's grapevine has spoken of ospreys in new places. There was a chat with a friend in a filling station. Her son had seen two ospreys in the air "apparently very distressed". There were other customers in the filling station and you cannot be too careful with such information. You never know who is listening. I needed to know where. She knew I needed to know where. "You know," she said, "just down below us."

"Thanks, I'll check it out," I said, knowing that "down below us" probably meant a quiet little backwater in the foothills of the Highland Edge where she and her family live. I was less than half an hour's drive from the place, the afternoon was free of commitments, so I drove straight to where I had in mind, parked some distance away, put on a top layer of dirty green jacket, overtrousers and wellies, and picked a discreet and trackless way through bog and woodland heading towards the water's edge.

A stand of Scots pines darkened the wood beyond the well-spaced birches where I walked. I paused to admire the trees for Scots pines are among my very favourite growing things. They wear the landscape's clothes and the shapes of clouds.

The tallest of the pines was partially obscured by closer, bushier trees, but my eye snagged on a straight edge that seemed to project outwards from the heart of the tree like a shelf. Trees the shape of clouds don't normally have shelves. I backtracked through the birches moving slowly and softly, avoiding the bare places where a flying osprey would see me, trying to find a clearer sightline to the shelf. It took several minutes of cautious manoeuvring until I could see the tree clearly through a light screen of birch leaves. The "shelf" proved to be a huge eyrie. As I focussed the binoculars, it heaved a ponderous brown and white shapelessness up out of its core like a time-lapse film of a monstrous mushroom. The thing turned then, and became an almost fully-fledged osprey chick. It lumbered to the rim of the eyrie, defecated hugely over the edge, then glared all around, a gargoyle made flesh and feathers.

For years now I have lingered by this little loch and admired its pines and thought it looked perfect for ospreys, while the ospreys had been elsewhere. Now though, as the ospreys spread further and further out from the heartland around the big loch in search of new territory, the obvious attractions of this smaller, quieter place had finally lured a breeding pair. I lingered long enough to satisfy myself that the adults were feeding the chick and retreated as carefully as I had arrived. Back at the car I saw the adult male carrying a fish in to the nest, but he had not caught it in the small loch by the pine tree. Instead he had made a round trip of four miles to fish the waters of the big loch by the badger wood.

Ospreys routinely do such things when the osprey-watcher judges that there would appear to be more straightforward options, but I have learned to accept that the priorities of a human mind and a wild creature's mind rarely coincide. Several times now I have watched an osprey catch a fish on the loch, rise up out of the water then begin a series of climbing zig-zags. The purpose is simply to gain height, because this is an osprey with an eyrie on the far side of the foothills, and on the edge of a different loch. But the loch by the eyrie is deep and not artificially stocked so the fishing will be trickier. The bird opts instead for a six-mile round trip that crosses a hill pass at 1100feet.

More difficult to understand is the example set by an osprey about 40 miles east of here, and which nests on an artificial platform on the edge of an artificial loch that accommodates an artificial trout fishery. There are all the fish it could ever wish for on its front doorstep. But every now and then the male disappears for hours then returns with a flounder. The nearest source of flounders is the Tay estuary, a round trip of about twelve miles, an extraordinary labour if the only object of the exercise is a varied diet. Or perhaps the bird winters by the sea. Or perhaps it hatched out in an eyrie close to the sea. But one way or another it has acquired a liking for the taste of salt in the back of its throat.

That same spring when I found the new nest, the grapevine had more news. Twenty miles further north, and close to the small cottage I was renting among the mountains of the Highland Edge, a local gamekeeper told me about an osprey he kept seeing crossing a hillside near his home and carrying a fish in the general direction of a plantation forest near mine. Twice in two years I had seen an osprey flying behind the cottage towards yet another loch in the west, and there had been sporadic reports of a fishing bird there, but the keeper's information suggested a bird coming into the area from the east. So I began scouring the forest with the binoculars from across the glen looking for suitable eyrie trees and anything that could be an osprey eyrie. I found nothing and I saw no osprey crossing the hill. And then, very late in the nesting season, the grapevine coughed up the answer.

The bar of the local hotel is a kind of unofficial focal point of much of the grapevine's dealings. One of the regulars who drives the hotel's minibus along the same route at the same time every day told me he kept seeing "a big white bird carrying sticks across the road". He didn't think it was a gull. Did I know what it could be? So I told him what I thought it could be and asked for precise details. Armed with these I climbed a hillside, found the nest, and understood at once why I had not seen it from across the glen. I had been scrutinising treetops. The nest was on a mobile phone mast.

By now the keeper's wife had also found it and reported it to the RSPB and the local police wildlife liaison officer. The nest was late and no eggs had been laid there, so probably a frustration eyrie built by a pair whose original nest had failed, or young birds trying to establish a first territory of their own. So it was decided to demolish the nest and erect an alternative artificial site nearby. There were two compelling reasons that assisted such a decision. One was that the nest would not have improved the signal from the mast in an area where mobile phone signals are imperfect to say the least. The other was that if word got out to the egg thief fraternity that ospreys had built an eyrie with a steel ladder up to it the birds' ambitions were doomed.

So on the threshold of one more spring I scan all my familiar shores and trees and hillsides and skies for the season's first glimpse of the prodigal's return. The wild year waits too, like the still-as-ice-loch awaiting the first breath of the advancing spring. The osprey's return is the first true sign of nature's commitment to the new season. Among the dozens of birds that now home in on their chosen treetops all across the Highland Edge I hope two in particular have a better idea in their heads than a mobile phone mast.

4
Hail and Farewell

SO THE OSPREY is back from Africa, home where it belongs in almost every quarter of the Highland Edge, and one ritual of spring here is in place. Almost at once a second ritual of equal importance to my wild year begins to unfold. On certain watersheets those Icelandic itinerants of our winter landscapes, the whooper swans, begin to muster in readiness for the long trek up the northern ocean. And here and there among these hills and lochs and lochans, Africa briefly eyes Iceland. By the end of April the ospreys are in place and the whooper swans have gone. You can never be sure, of course, that when a herd of whoopers takes off from an accustomed loch in the morning that this is a final leave-taking, or merely one more foraging trip to the fields and shallow waters…

…Three whooper swans still lingered on the leeward shore of a mountain loch. They had been there an idle week when I would have expected them to be on the move up the west coast to the renowned gathering places in the north of mainland Scotland and in the islands, where the big skeins take off for Iceland and the nesting grounds.

Near the three lingerers, two mute swans had begun to look as if they might nest in a place where I have not known mute swans to nest before. So on a late April evening of blue-white skies tending to purple in the west, as the sun sank behind Balquhidder's mountains, I broke a long writing shift to check the birds' progress.

I was disappointed to find the mute swans too far down the glen, several miles from where I had seen mating overtures two days before, and I almost turned back at that point. But the beauty of the evening tempted me on, that and a sixth sense that over the years of my wildlife watching I have learned to trust utterly.

At the mountain loch the three whoopers were where I had last seen them. But there was also a strange, broken light on the water in the middle of the loch. Sometimes, if you go often enough and alone into wild places, you are touched by an intuition concerning the most apparently ordinary things, like a patch of broken light in the middle of a loch. This one struck me as mildly curious, though for no obvious reason. So I made a long and quiet downhill detour among scattered trees to close in discreetly on the place.

A few yards from the shore I began to hear voices, many voices. They grew loud. Mountain rocks echoed the tumult and threw it back across the open water in a thrilling confusion. I know these voices. They put Arctic hints on my native landscape. They layer my favourite places with embellishments of far northern wildness. No sound I know gladdens my world or restores bruised spirits quite so powerfully and instantly as the opened throats of whooper swans.

But these were not the three birds on the far shore, for I could see them clearly, and they were heads-down-tails-up in the shallows. I came to a big rock and peered over. A hundred yards away, my patch of strange broken light materialised into a tightly packed and thoroughly agitated band of twenty-three whooper swans. The noise was astonishing. Every element of the considerable vocabulary of these swans was on the air, and so was one I had never heard before, a loud, stacatto purr.

The birds were on the move. They coursed across the loch in a wide circle that took them close to the three disinterested feeders on the far shore, and there the horde paused, calling and calling. I never saw such a blatant invitation before in all nature. But there was no response.

So the horde ploughed on, another vigorous, swimming circle that brought them closer to my shore. I called out to them in my poor imitation of one of their calls. In the right circumstances, whooper swans will swim towards the call, answering back. These were clearly not the right circumstances. They barged past, tall-necked and bow-waved, and the mountain loch rocked under their passage. Their wake slapped ashore against the stones at my feet.

They closed in again on the three. The same invitation met with the same indifference.

I was convinced that all this was a prelude to take-off, convinced that at any moment the palpable tension among the birds would snap and the whole flock would fly. One bird lunged forward on threshing wings, and was airborne for a few low yards, but none followed, so it landed again and turned and swam back to the flock. In the glasses, it wore the grayish plumage of a young bird.

Any second now, I told myself, any second now…but an hour and a half later, nothing had changed, except that the light had faded into dusk and the birds had grown more agitated, the calls more frantic, the echoes more dizzying.

On one more circuit, they paused beside the three feeders and one of the three was finally persuaded and swam out to join the throng. But the other two fed on. The dusk darkened, the flock powered on round and round the loch. In the west, between the mountains, the purple had paled to pink and a crescent moon brightened there.

I decided to climb back up to the road in the little that remained of the light. In the unlikely event that the flock would fly at such a late hour, I could watch from up there.

I reached the road, turned and looked back, and the birds flew. They flew in a broken mass of stretched necks and powering wings, low and slow at first, rising a dozen feet in a quarter of a mile. Then they banked in a long curve and began to circle the loch, flying directly above the two swans they had left behind. The pair looked up as they flew over, but nothing more.

Twenty-four swans, calling and calling as they flew, circled the loch again, gaining another dozen feet, then again, then again, then again, until, by the fifth circuit they were flying at my eye level. Then they started to fly close into the mountainside across the loch, and there, riding the thermals, the climbing steepened perceptibly. The swans were climbing the mountain.

On perhaps the tenth circuit, I saw them for the first time against clear sky above the mountain ridge, and flying at about 1500 feet.

After two more circuits, the whole flock suddenly levelled out and pointed west. They were now at perhaps 1800 feet, and suddenly there was a clear unanimous purpose…except that just as suddenly, it was shown to be less than unanimous. Five of the birds began to glide with their wings rigid and down-curved, and from 1800feet, these five fell in an unbroken sky-dive back to the surface of the loch. I never saw a swan dive before. Necks dead straight, wings curved and angled back, the dive all but vertical until the last few feet when they pulled out in unison and settled at once onto the water.

So nineteen whooper swans flew west, between mountain silhouettes, against a blue-white sky banded with yellow and gray-ish pink and lit by a crescent moon, and I watched them until they simply faded into distance beyond mortal sight, and by then their calls were out of earshot too.

Down on the loch, the five swam to the shore where the other two had never stirred, and all seven stepped from the water onto the shingle shore, and prepared to settle down for the night. By then I was cold and had developed roosting notions of my own.

On the way home I passed the mute swan pair again, close in to the road and close together, and my notion that they might yet attempt to nest in the glen was restored.

In the thirty years or so that swans have been swimming and flying into my life, I have learned a lot about their ways, and for every new thing I learn I have a hundred new unanswered questions. And still, all I know for certain after these thirty years is that I do not think like a swan.

At noon the following day, two whooper swans stilled lingered on the leeward side of a mountain loch. They had been there a week and a day.

The incident had an eerie echo eleven months later. I was on the shore of the same loch a few days after a heavy late snowfall. The loch was quiet and flat calm until a relaxed herd of sixteen whooper swans at its west end suddenly galvanised and grew loud. They bunched up behind a single dominant bird. The tranquil swimming and feeding and preening was over, and now they swam together and at speed, but in every possible direction. As the sound grew, so did the visible prelude to take-off, a sustained bout of head and neck pumping. I found a vantage point, settled the glasses on a fencepost and watched.

The world was black and white. The mountains were wreathed in snow, the ridges piled and underscored into cornices. The forested lower slopes showed up in the low light as a blackish-bottle-green, and their reflection in the loch translated that into a profound black. It was against that blackness that the vivid white of the birds now swam. By now they had formed into an organised vee behind the leader, and now they headed out into open water, and now they flew. Swans never work harder than when they are trying to become airborne. Necks are straight and low and parallel to the water. Wings are hugely raised then lowered to water level. This raises the birds up off the water, but only to the height at which they can thrash it with their feet. A dozen wingbeats is usually enough, and then they fly a couple of feet above the surface, but instantly, on that day of the big snow and the black water, they were joined by their upturned flying reflections, so thirty-two swan shapes flew in a two-tiered vee.

But this time they flew east, climbing gently as they flew, until they were high enough to clear the trees at the east end of the loch. By then their reflections had drowned in deep water and they flew on unaccompanied. Two days later I saw what was almost certainly the same group on another loch five miles away.

I often wonder about the restlessness of whoopers on their winter migration, and it may simply be that they are away from home, far from their Icelandic nesting grounds, and that their more or less constant travelling among known watersheets is no more than a way of passing time until they can return to where they want to be. Some whoopers don't migrate and thole the Icelandic winter often with the help of handouts from people, for the bonds are very close between Icelanders and whooper swans. Others – a small but increasing number – find conditions on migration to their liking and try to breed here, especially in Orkney and Shetland, occasionally in the Western Isles, even more occasionally and almost inexplicably, in Highland Perthshire. There are late stopovers on the mountain loch almost every year, and every year I invite them to think about staying. It hasn't happened yet, but a few more late snowfalls like that one and who knows? One consequence of climate change might well be colder winters in Highland Scotland as the benevolent effects of the Gulf Stream are offset by colder water from melting Arctic icecaps and glaciers, in which case Scotland may start to look more like Iceland. The loch is available and the swans already know the way there.

5
Soliloquy by an Empty Reed Bed

EARLY JUNE and the day vivid, the sky a gracious blue, well patched with high, gray-white convoys of clouds shunting loosely north-eastwards on a pugnacious breeze, an eddy of which licked at an edge of my writing page and made it curl up. I leaned my left thumb across the top of the page for a moment until the eddy passed on. I was sitting by an empty reed bed, a quiet bay at one end of the loch that is the beating heart of my territory. The reed bed was not empty in the way that a holed trawler net or a defeated whisky bottle is empty, but it was empty of swans, and swans have been my four-season purpose by the reed bed for the better part of half my lifetime, and just occasionally, the worse part.

All things being equal, on such an early June day as this there should be a huge nest in the reed bed. The biggest I remember was eight feet across, four feet deep, anchored to the rooty floor of the reed bed by its own weight. Its eggs should be close to hatching, late by Lowland standards but timed so that the cygnets are spilled from the eggs into Highland Perthsire's brief season of plenty. The nest should be well screened by the burgeoning reeds, and the lethal season of floods should be over.

All things being equal.

These pencilled words on the page in my lap mock the reality of what my eyes see when they lift to the landscape in front of me, for this is the most unequal swan territory I have ever known. It was never more unequal than this spring of which I write for it has treated the swans with unprecedented cruelty, even by the perverse standards of the place.

Cruelty?

It is not a word I like to apply to nature's dealings with its own tribes. Surely cruelty is inflicted on swans by people, not by nature? Ask the staff at any one of the country's swan hospitals. Hear stories to foster contempt for your fellow mortal; feel it rise up and stick in your craw like a fishing hook in a swan's gullet, an airgun pellet in a swan's head, a pair of crossbow bolts that crucified a swan to a barn door by the wings. My love of swans has put me in touch with people who verified all those stories. But cruelty at the hands of nature? Mother Nature with the blood of swans on her hands…?

…Each year, the reed bed spreads. I would guess that it has doubled its area in the last twenty years. An old-timer in the village a mile further north remembered when the reed bed was a potato field, a fragment of a small farm, an extreme manifestation of that species of agriculture known hereabouts as "marginal", a margin of the margins. When it stopped being worked and drained, nature barged in. Now a regime of winter and spring floods has been much assisted by plantation forestry on the surrounding mountainsides and an enforced end to the local practice of dredging the river for gravel where it flows into and out of the loch. All that has edged the reeds deeper into the shoreline grasses, because the shorelines are more water and bog than they used to be. Thus bolstered on the shoreward side, the reeds' irresistible system of powerful underground roots and stems wades ever further out into the shallow water of the bay.

On this June day, the reeds are dark green and three feet clear of the water. In another month they will reach seven or eight feet, they will thicken and their colour will grow rich and gold-tipped. When a sunlit breeze goes through such a reed bed it rustles and sways with a fluid grace and the lit gold and bottle green is a fit garb for a swan nursery. The best of swan summers here are electrifying. But the best of swan summers are rare events. This year, summer's promise will be unfulfilled again, and the reed bed is host to that emptiness I dread. I have grown well accustomed to it over the swan-watching years, for mostly this place does not work as a swan nursery. The failure level of eggs and the mortality of cygnets are dire. There were six successive years when no cygnet fledged at all. Six years, fifteen nests, around seventy eggs, seven cygnets hatched, not one cygnet fledged, a colossal endeavour of persistence by the birds in the face of hopelessly overpowering natural odds, and utterly futile. Even if the seven had fledged, the mortality rate among first-year swans is so high that it would still have counted for almost nothing.

And now I feel for the first time in all my years here as if something has gone from the place, a symbolic, living thing representative of all swans. The solitary hope I nurture for the next twelve months is that I am wrong. I try to keep my ambitions for these swans modest.

In my mind, that absence is a flame, a cool flame, white like a single swan's wing. Once I saw a huge swan raise one wing towards me while it preened. Low sunlight shone clear through the wing, illuminating it from behind so that it seemed to be lit from within, lit like a flame. And when the breeze lifted and disordered some of the wing feathers, the flame danced. The image is among the most durable of all my swan-watching years. I see swan wings in my hearth; and the white-wing-flame stands in my mind for all the swans of my life and all my thousands of swan days all over Scotland and occasionally on other shores from Iceland to Alaska, but mostly on this loch shore, by this reed bed.

And on that June day, the reed bed was empty and I could not sense the flame. I wrote in the notebook on my lap:

"Perhaps there remains a spark that a benevolent air may blow kindly on and rekindle. But I can't detect it, and there are only contrary winds at work. Suddenly, even as the reed bed begins to burgeon, it feels like a dead place because it has failed its swans. And even though the little grebes still cry their unearthly chuckles unseen in its maze of tiny channels; and even though herons stake out its fringes; and even though swallows and martins cling to the reeds in their hundreds whenever a squall shivers up the loch and through the bay; and even though the otters plough sinuous furrows through the grasses between the bay and the river, what I see mostly is the swan-emptiness."

What still puzzles me is why the swans have persisted here for so long. I am, of course, grateful that they do, for they have heaped raw material on my writing desk, but I am also baffled. What is it that commands their loyalty? And for that matter, why do they demand such a huge territory – about half a square mile of water – when the difficulties of maintaining even a small territory here would be quite beyond what most mute swans would endure? Conditions are so routinely hostile to what they set out to achieve each spring. Yet these birds, and the huge old cob in particular, are extraordinary creatures even within the extraordinary world of swans. It seems never to occur to them – to him, for he is the constant force of at least the last 25 years – that there are easier ways to make a living, easier waters to call home. I have only ever known him to be away from the loch for a handful of days at a time, never in the nesting season, always in extremes of circumstance. These have included winter freezes, floods so high and widespread that feeding became impossible. And once, when the death of a mate left him alone, he went off within days to seek a replacement. He was back in two weeks with a new mate, and how I wish I could have seen him go, followed and watched. How did he know where to go to find a mate – the waters of his own young years? How did he go about luring a stranger swan to leave its familiar surroundings and fly with him to a landscape such as few mute swans have ever seen, for this has been an isolated mute swan outpost, and all the nearest populations of swans are on the Lowland side of the Highland Edge, beyond the mountains?

However he resolved those problems, his actions demonstrated the work of a reasoning mind. He was not prepared to grow old alone or await the chance arrival of a passing vagrant. Writing it all off as "instinct" won't do. True, a kind of survival instinct is the driving force that controls the lives of most wild creatures, but it is unwise to assume that it is the only force at work. If that were true, all swans would behave in the same way, so would all eagles, all otters…but individuality is pronounced in many creatures, especially if they live long and isolated lives rather than short and sociable ones. I have met clever swans and stupid ones, confiding ones and aloof ones, bold ones and timid ones, swans that accept their lot and swans that take decisions to make things happen. And there are swans like these, swans that conceal more than they reveal, swans that baffle me more than they enlighten me.

And it did not occur to the cob to settle for quieter waters with his new mate. Instead he brought her back to the only territory he has known as a breeding bird, to this out-on-a-limb corner of a Highland loch. Perhaps it is as simple as that explanation I was handed about why the badgers returned to the woodland sett after it had been cold for twenty years – once a badger sett, always a badger sett. Once a swan territory, always a swan territory. Or does the swan (the badger for that matter) know about what we would describe as "a sense of place". Does landscape matter to the swans, or at least to that cob in particular, the way it matters to me? Are there swans that simply prefer Highland landscapes to Lowland ones, the unfettered scope of a loch to the confines of a pond, at a remove from human beings rather than in their midst? If so, do they make conscious decisions that indulge their individual preferences? In general, nature's interest is the wellbeing of the tribe, not the individual. Golden eagles or ptarmigan *need* a rarefied mountain realm or an isolated island one. Mute swans do not. Indeed they prosper in the Lowlands. By far the most fecund mute swan nest site I have ever encountered is an old mill pond on a West Lothian farm, a dozen miles from Edinburgh, a landscape whose every detail is man-made, right down to the rocky little island where they nest. Their benevolent human hosts cater to their every need and they prosper. In the forty years since the island was built, around three hundred cygnets have flown from that pond. In the same period, I doubt if the Highland Edge reed bed has fledged thirty. In the six years when they raised no cygnets here at all, the Lowland pair raised forty. But still the birds persist, and still I bear witness to their struggles in a landscape more suited to eagles.

And I would have to admit that the unequalness of it all, the colossal and sustained nature of the birds' struggle against other forces of nature, is the most compelling aspect of the situation for me, the reason why *I* keep coming back for more. I want to know how far these birds will be pushed, how much they will thole. And I want to witness – to *share* if I am honest – their small and occasional triumphs over such adversity. Because every now and again there are moments when it appears as if they have let me become a part of their landscape. Something happens and they place me in a position of trust. These I hold among the most precious moments of my working life, for it is then that a key turns and nature beckons me into its inner circle. Briefly I can see with other eyes and there is no higher prize.

That is the way it has been by this reed bed these last thirty-something years. And if I am troubled now by the emptiness of the reed bed, I should also set against that emptiness the knowledge that these swans have known many dark hours and have recovered from them all the next season or the one after; that they move to nature's timescale rather than mine; that just because this is the nadir of my own swan watching years does not necessarily mean it is a nadir for the swans.

So the one thing I will not do now is turn my back on the place. Perhaps this year of the emptiness has purpose. Even while the swans are gone from the centrepiece of their world, the centrepiece itself remains, and the swan-emptiness must be watched over and pondered, for it is part of the story too. Once a swan territory always a swan territory.

It seems such a long time since my involvement here began, but in the story of the loch, the river that feeds it, and the mountain landscape that holds them both in place, it is less than nothing at all. My labours here are of no more consequence to them than they are to the swans. The pencil fills the page in my lap because it satisfies not the needs of mountain and swan, but needs of my own. This is what I do. This landscape of mountain and forest and river and loch is mostly where I do it, my nature writer's territory. And these pages are what I make of it, a self-portrait with swans.

The little breeze died, the corner of the page lay flat again, and I returned my left thumb to the margin of the page.

6
The Tainted Miracle

TO SPEND MORE than half a lifetime watching a pair of mute swans must seem to many a stranger something between pathetic obsession and a spectacular waste of time, even for a nature writer. This is, after all, a landscape of golden eagles and ospreys, otters and badgers, wild cats and pine martens, red and roe deer, mountains, forests, rivers, dozens of lochs beyond this one; there is so much to choose from. All these have occupied me to a greater or lesser extent, but none has laid claim to me, to my work and to my emotions and instincts like the swans. My early work here culminated in a book called *Waters of the Wild Swan* (Cape, 1992) which led to a BBC TV programme, *Addicted to Swans.* That in turn led to a series of Radio 4 broadcasts from the loch and a subsequent little book, *The Company of Swans* (Harvill 1997). My first novel, *The Mountain of Light* (Whittles 2003), was set in a Highland Edge landscape, a modern myth woven around the old swan legends and which evolved directly out of the swan watching years. I have lost count of the number of newspaper and magazine articles, poems and stories I have written about swans in general and these swans in particular, and of how many hundreds of photographs I have taken. And still I cannot be away from this shore for long, and still these birds exert an influence on me unlike anything else I have known in nature.

I could defend myself. Only three of my twenty-two books are swan books, and then the case for the defence weakens: there are swan references in almost all my non-swan books. Bookshops and publishers and the arts establishment divide prose writing (none-too-comfortably for my liking) into fiction and non-fiction. My own writing divides into swan and non-swan, and that is none-too-comfortable a distinction either. And I believe it is neither a pathetic obsession nor a waste of time that has occupied so much of my working life, but rather that I am something of a throwback, prey to an ancient process I do not wholly understand, for human beings have been fascinated by swans for 20,000 years. In these last thirty-something of the 20,000, my relationship with the reed bed birds has changed from casual observer to a kind of fretting, exulting foster-watcher, as attentive as I am uninvited and unappointed.

Year on year I begin again, bringing all my accumulated knowledge to bear on the new Highland spring. Yet it is never enough. Year on year, nature contrives new ordeals for the swans, new mysteries for the swan watcher. The more I learn about them, the more I realise how little I really know, how much more they know than I do about themselves and the landscape we share. I try and separate what is instinct in the bird and what is intellect. And I try and reappraise a few of the hundreds of stories that people have woven around swans. Some of them will be 20,000 years old and God knows who the story-tellers were. Others are less than twenty years old and these stories are mine.

There is, for example, the year of the miracle and the direct consequences that so tainted it. The photographs I took seemed to be irrefutable evidence that nature had decided to bless a union of swans it had become ritually accustomed to curse. For once the spring was wholly benign. The birds built their first nest in early April and in warm sunshine. The pen sat early by her own standards. Spring on the Highland Edge almost invariably flourishes a sting in its tail, but that year there was none; no floods, no gales, no late frosts or sleets or snows. In the first week in June she responded to my familiar, ritualised presence in an unfamiliar manner. She was near the nest on the water and I thought at first she was simply enjoying a brief respite from the boredom of the long incubation routine. Normally she would swim over to greet me. Once she regarded me placidly from a few yards away while she permitted me to count the eggs on the nest. A few moments later she climbed calmly back onto the nest, rearranged her feathers, removed the reeds and down with which she had covered the eggs, and sat while I talked to her and photographed her. But this time, she scuttled out of the reed bed in a state of some agitation. Only when I saw her in clear water did I realise that she was towing a single file of five very young cygnets behind her. When she was out of sight I sloshed out to the nest to see what she might have left behind. The answer was nothing. She had laid five eggs on her first nest and hatched them all. If it had happened on the West Lothian farm pond it would have been regarded as something of a failure. But in my long experience of the reed bed nest site it was unprecedented and cause for celebration. I took the travel-scarred little hip flask from my backpack and tilted it first towards the swans, then towards the waters of the nest bay, the loch and its fringing hillsides, and drank a taste of something peaty and Hebridean to the health and long life of the new swans.

The most dangerous parts of a swan's life are its first few weeks, then its first winter. The parent birds can be formidable in defence of their cygnets. I have seen them fend off dogs and foxes and herons and otters, but they have no defence against the pike that pulls one under and swallows it whole, or against the Highland weather. They are also not above abandoning or drowning an obvious weakling in the brood. But that summer, these five prospered, and within a few weeks the adult birds had reinstated their confidence in me and the cygnets learned it by example.

There was a day of late September, the swans together on a spit of land far down the lochside. I went to the old familiar corner of the shore of the nest bay and through a warm noon hour I enjoyed the sun and the familiarity of the landscape and its wildlife. From time to time I turned the glasses on the swans. They too were at rest, although the big cob was on his feet and watchful. They were 500–600 yards away. The water was still, the hour windless. Sound travels marvellously under such circumstances. On a whim, I stood up and called to them.

"Come on swans, come on! Come on swans, come on!"

I have done this for years and the response is almost unfailing. As long as the birds are within earshot – and not necessarily even within sight – they stop what they are doing and swim to my shore. That moment of response ignites a small thrill in me. I have spun a thread of communication out across the water to the birds. They in turn spin out a direct and unambiguous response. The expectation of a handful of food is undoubtedly part of it, but at that time of the year they are surrounded by food in the loch's shallow bay, with no need to travel anywhere. I choose to believe, then, that part of it is also trust in my presence on the shore, and that at the very least they regard me as a benevolent fragment of their landscape. So, there was a day of late September and I called…

…I saw the cob straighten and turn his head. Then he stepped into the water. The pen looked up from her spine, which is where she happened to have laid her head and neck. She uncoiled and stood, and stepped after her mate. And one by one the cygnets did the same and formed up, well spaced out, behind their parents. A procession unfolded up the loch and it seemed at that moment that it had been orchestrated entirely for my benefit. This, it seemed to say, is for you. The straight line never wavered, the formation held and they came on up the loch until they were all around me on the shore and in the shallows and I felt briefly a part of it all, a part of the pageant of that place, a wild fragment of it, like a passing raven or a knee-deep heron or an eagle crossing a watershed or the very hills themselves. Most of the time nature works hard at keeping people at arm's length and for good reason, but if you go often enough and respectfully enough and watch and learn and try to understand, there will be moments when it nods to you and lets you in. You cherish such moments for they are as priceless as they are rare and brief. They are your reward, and they are all the reward there is.

And then, with the swans all around and the sun on my back and the day blissfully at peace with itself, a cold shadow of premonition troubled me. I am essentially optimistic in character, but something stole across me then that was born of everything I have ever learned about this nest site, these swans, the knowledge that nothing comes easy here. Yes, the spring and summer had been perfect and the swans had prospered. That September hour was as close to my own definition of the word "sacred" as any hour I have spent on that lochside. But suddenly I sensed that sooner or later there would be a reckoning. It would prove to be later rather than sooner, but there was a reckoning.

The birds were fully fledged in mid-October, they wintered well, and at the beginning of the following spring all seven swans were still together. Cometh the hour, cometh the reckoning.

First there was the weather. It was as vicious and unrelenting as the previous spring had been sweet and tranquil. For weeks storms battered up the loch and drowned the old season's reed bed. The swans' first two nests and a dozen eggs succumbed. But it also became clear that the pen was working very much on her own. The cob's insistence on half-a-square-mile of territory was working against him. On smaller, less hostile waters than this one, it is common enough for cygnets to make their own way into the world at the end of autumn or the end of winter. If they are still around when the new nesting season begins they are driven out by the cob, sometimes violently. By mid April, there were still seven swans at the north end of the loch and that was five too many. If the cob had simply been content to patrol the nest bay the outcome might have been different. The five cygnets were mostly together in the other bay beyond the river, two hundred yards from the nest in a straight line but two or three times that distance if you have to swim out round the mouth of the river, which is how the cob chooses to patrol his territory. Hour after hour and day after day, I watched the cob hound his offspring of the year before around that far bay, driving them one or two at a time to the mouth of the bay and out onto the open loch. And as soon as he turned back for the others, the expelled ones would swim back to where they had been half an hour before. The cob was quite incapable of driving them all out of his territory. This exasperating behaviour persisted for weeks, and then for months. It became his daily and day-long ritual from April to September. His energy and purpose never flagged, and the young birds never left. He never tried a different tactic. He never tried to defend a smaller territory. They were in his territory and for reasons known only to himself, he needed it all.

Back at the nest, the female did not see her mate for hours at a time. I arrived one late May morning after a night of unbroken downpour to find the reedbed underwater again. The only scrap of anything like land that showed above the water was the top three or four inches of the nest, and there the pen sat on, and I knew that the third nest was doomed too. Late in the day she left it and its four eggs. The cob was far across the loch, chasing cygnets.

I left them for a few days. The waters fell back, the reedbed slowly re-emerged, the pen fed and recovered some of her energy and the cob expended all of his in pursuit of the unattainable. When I returned it was to find the pen alone on the lochan to the north of the reed bed. During the floods, the narrow, shallow, rocky channel between the lochan and the main loch widens and deepens and becomes easily navigable for a swimming swan. The pen had swum that channel and on the lochan's boggy shore she was building a fourth nest. I leaned against an alder tree a hundred yards away and watched in disbelief. It was early June now, and if she did finish the nest, and if she did lay eggs there again, and if the cob did abandon his pursuit of the five young birds long enough to mate with her again, and if the weather did relent, and if she sat tight for another five weeks, and if the eggs did hatch, it would be mid-July, and the chances of the cygnets fledging were rather less than non-existent. It is inconceivable she was unaware of all that, yet she was still driven to try. And what happened was this:

She did finish the nest. She did lay eggs in it. She did sit on them. But I never once saw the cob on the lochan, and it could be that that June they lived out utterly separate existences. When the rain returned in late June, by which time she had been sitting for three weeks on her fourth nest of the season, she finally admitted defeat. She left the nest and the lochan and swam back into the reed bed bay and the purpose of her wild year was over.

But after another month of rains and gales and generally disgusting weather, her fourth nest was still intact, still perfectly serviceable, its four cold eggs still in the shallow cup she had made for them. I have often thought that if only the swans would opt for the lochan as their territory and nest there rather than in the reed bed, their chances of success would multiply. But they never do and I never know why and I repeat to myself the only certainty that all the swan-watching years have taught me: I don't think like a swan.

That autumn and winter there were seven swans on the loch again. The hostilities ceased with the autumn. A family of eight whooper swans arrived on their migration from Iceland – two adults and six youngsters, a rare sight. And then suddenly more mute swans began to arrive from the south. I believe that swans in flight respond eagerly to the sight of swans on the ground or on water. Firstly, the vivid whiteness of a group of swans is conspicuous at a great distance and will catch the eye of the flying birds. Secondly, the presence of a group of swans on the ground or on water implies the availability of feeding, and that too attracts the flying birds. Over a few December days eight more mute swans arrived and four more whoopers, so suddenly the loch at the heart of my own territory was host to twenty-five swans, and I began to wonder if these two consecutive breeding seasons of such conflicting fortunes might have the accidental effect of creating a winter haven for swans. My mind's eye dallied with the prospect of hundreds of swans in the bays and the lochan, of establishing a low-profile local nature reserve to keep the place free from disturbance and give nature its head.

The whoopers moved on, the way they do, for they are restlessly nomadic outwith their breeding season. The mute swan numbers fluctuated, but the five young birds remained and a sixth solitary bird lingered. So the following spring developed a similar pattern to the last one with similar results, except that in August the big cob disappeared. At once, swans poured into the nest bay, their incursions unrepelled. The pen, his mate of four years, was powerless to stop them. Two weeks before, she had turned her back on her second nest. Almost immediately she had gone into moult. The long ordeal of the nesting season – fifteen weeks, two nests, ten eggs, no cygnets – had left her weak.

Throughout her time on the second nest, from early June to mid July, her mate belatedly recognised the changed circumstances caused by the continued presence of the five young birds and rarely left the mouth of the nest bay. As long as he patrolled there, none of the other ten swans on the loch risked his splendid wrath. It was as if he was suddenly conserving and concentrating his energies. The pen, screened on a nest of reeds by the swaying, gleaming, whispering curtains of the reed bed itself, had waited a month after ritual spring floods had washed out the first nest and scattered its six eggs. I had watched those floods grow, watched loch and river rise as the rain fell and fell, watched the swans build the nest walls higher and higher and higher – more and more broken fragments of old dead reeds laboriously gathered and painstakingly shaped into a perfect, massive circle. Once again the nest and the sitting bird were all that showed above the water level. As long as the weather remained windless, and as long as the swans kept shoring up the nest walls, there was a chance. But after a week of all-but-ceaseless rain a south-westerly gale built up walls of waves, and these finished it. Mostly, mute swans do not build again when they lose a nest or a clutch of eggs. But the tradition of this nest site – that cob and his various mates – is otherwise. Mostly they do try again, and again, and I still harbour the memories and bear the scars of the year when the birds responded to a succession of floods by building five nests in a single season, laying 26 eggs, but hatching none.

Now, a nest had been abandoned, and I watched and waited, and watched and waited, and nothing happened, except that the nesting season grew late, and late hatchings for mute swans are almost invariably failures.

On June 3, the pen was sitting again. I guessed four eggs. The weather held, the cob remained attentive and close, and an unlikely late hatching suddenly looked possible. I judged that if she was going to hatch anything at all, it would be around July 20. But on July 20, she covered the eggs and swam out through the narrow channel the width of one swimming swan that the two birds had made through the reeds to the bay, and began to feed. She did not go back to the nest. Two days later, I waded out to the reed bed to see what she had left there, looking for explanations, straws to clutch, anything that might assist my long preoccupation with the mind of a swan, for it is a perplexing thing. The two birds watched from a hundred yards away.

It is a strange terrain for a walking human. To begin with there is the surprise of stepping in among the first reeds from the squelchy mud of the shallows. You step *up*, onto the rooty floor of the reed bed and the hidden land is immediately firm under your feet. Then, you realise that you can see nothing at all other than reeds. In midsummer they are seven or eight feet tall, and they mutter all around you in the breeze, and sway restlessly. And regardless of how carefully you thought you had marked the position of the nest from the hillside across the loch, it is not where you thought you had placed it. But then, you are not where you though you would be either. So you part the constantly reforming screen of the reeds in front of you and step forward, and on you go.

Suddenly there is a space, a clearing more or less circular, perhaps a dozen feet in diameter. The clearing has been created by the swans felling reeds, biting through them at the base, thousands of reeds, weeks and weeks of primitive labour. Each reed is thrown over the swan's back by a twist of its neck and onto the water. Often it needs to be picked up and thrown again, sometimes twice more, before it can be added to the heap that becomes the nest.

The top of the nest is level. The eggs are in a cup lined with down plucked from the swan's own breast, but they have been so well covered that you have to feel around the centre of the nest until your careful fingertips detect the hard outline of a shell. You pull away the covering of reeds, and count five eggs. They are quite cold. You cover them again and retreat, sadly.

And then the cob disappeared and the pen fled down the loch while the mob of unattached swans cavorted in the empty territory. I did not see the cob again but in November I had a phone call from a friend telling me that a swan had been found dead by a pond half a mile north of the reed bed. It had a big growth under one wing, and had been dead for some time. He was nearer 30 years old than 25, an astonishing survival in such a wild landscape, and the biggest mute swan I have ever seen. That confirmation of his death seemed to mark the end of the story, except that by the time he died, there were 20 swans on the loch, and these were one quite unexpected consequence of the miracle spring of two years before. Twenty swans, and a vacant territory, and of course there is no such thing as the end of the story.

The following year there was a pair on the territory, and one of them was almost certainly a native of that bay, and just possibly too young to breed. They did not build a nest. Last year, their second as a pair, they built four nests and lost them all. And so it all begins again and they begin to learn something of the fate that nature has handed them. I do not think the pen is the old mate of the big cob. Yet without hesitation they chose the historic nest site. It could be, of course, that one of them was hatched out here and has never left. Or it could simply be that the old saying is true – once a swan nest site always a nest site. And now as I write, in the autumn of their third year together, they and their four cygnets – hatched from their first nest during a wonderful spring and summer – are together, vigorously ready for winter.

7
The Kite's Tale

AT LAST, THERE, low over the hill, a red kite against the sky, going slow, too slow for the comfort of a processional wake of irritated rooks and jackdaws. The crow tribes prefer a bit of spice about their perceived enemies when they try and rough them up, a few aerobatics, a show of aggression, the occasional back-flip, presented talons, a turn of speed to test theirs. But not this; this slow-lane dawdler ignoring their every dip and dive, unswerving, undeviating, unmoved by every black oath.

Excitement surged through the hide though. Fingers pointed, binoculars glared, telescopes swivelled. The word "kite" escaped from everybody's lips, and that was a lot of lips. The hide was too full, and only the lucky ones seated at the front had a clear view. Most of the standing ones had to bend backs or crane necks or both, a posture that does not make for steady binoculars. Someone swore as someone else jostled his telescope tripod, covered his mouth with his hand, remembering where he was, trying to keep it down. Voice levels had been muted for most of the last two hours, the conversation a kind of forced good humour, but close to the surface lay a muffled tension fuelled by an ill-tempered January northerly with snow on its breath that blew straight into the hide through glassless windows. So the hide was too full and too cold. And suddenly there, low over the hill, was a red kite against the sky, going slow, and voice levels rose.

The sign in the village post office had said "Red Kite Open Day". The concept had instantly appalled me. The wording made me think of show houses. Then I thought of show kites. Then I grew angry. I thought:

Why do people do this to wildness?

Why must they smear it in PR gloss and show it off?

What on earth were they trying to prove at a red kite open day?

And then I thought I would go and see. I got there early, so I was one of the ones with a seat by the open window and a clear view. I settled quietly for a long wait which is one of the things I do best, but as I waited I feared the worst and it was to prove far worse than I feared.

The hide is on a farm. I met the farmer by the car park. He is a delightful man who thinks hard about his job and works his land well. I admired his attitude, his values, his priorities, his eagerness to give nature its place on the land he works. I have no argument with him at all.

A hundred young red kites imported from mainland Europe have been released near here over ten years. The site, immediately to the south of the Highland Edge, was one of several across Scotland carefully chosen in a thoughtful and wholly praiseworthy attempt to re-establish the species in its old homelands from which the Victorians had extinguished it. If you read only the statistics you would have to conclude that the project was a success. If you witnessed the consequences of a Red Kite Open Day you might conclude that those elements of the professional conservation establishment behind the project had taken leave of their senses.

The kite low over the hill against the sky going slowly and diffusing the corvids' hostility by its relentless slowness was headed for a stand of half a dozen big trees perhaps 200 yards away. There it perched, and although its shape was broken up by the winter-bare branches, its presence in the tree was betrayed by a vivid red wing-tag in its left wing. The colour denotes a bird from central Scotland. A different coloured tag in the right wing denotes the year it hatched. The same system is used by all the release sites. So with good binoculars, and often with no binoculars at all, or if you find a dead bird in your travels, you can tell where it has come from and how old it is. How valuable the information will be to your future wellbeing or to the wellbeing of kites is less than clear to me.

In the hide, a great deal was being made of the wing tags. There were posters that explained the colour coding, and the man from the RSPB explained over and over again to new arrivals how the system worked, and he had wing tags in his hand to show them. He was everything a front man for such a project should be. He talked knowledgeably, he was endlessly patient, he was as good with children as with the seasoned veterans carrying lunch in old canvas rucksacks. I have no argument with him either. Well, yes I have. He was an enthusiastic champion of the tagging system, of the importance of the scientific experiment. For him, the science of it all justified everything, and for me that has never been a good enough reason for meddling with the wildness of things.

Another kite, another hubbub. Then another and another, and soon the trees were hosting twenty red kites, and you could count them in the branches by counting red wing tags. The conversation grew animated as people tried to spot the colour of the left wing tag and then consulted the wallcharts and shouted its age.

The philosophy of the place is this. Red kites congregate in the winter close to the release site, and roost communally in a wood. If you know where the wood is you can watch in the late afternoon as the birds home in from every compass point. They come close and they fly slowly so the opportunities for birdwatching are extraordinary. Then they perform a stylised sky dance en masse before they settle for the night, and that is when you realise that the red kite knows every flying trick in the book and a few the book doesn't know about.

It was decided to ease the birds through the winter by creating a feeding station, putting out scraps of meat every afternoon, for the kite is a scavenger to trade. And by putting a hide near the feeding areas, people would be able to see kites at close quarters for a modest charge. And then the idea took wings of its own. Down on the main trunk road a couple of miles away, new signs appeared pointing the way to the hide. And to coincide with road signs the open day was announced, and the kite viewing would be free. Other attractions were soup and burgers in a barn beside the car park.

The hide quickly filled. Then the car park filled. Then the hide and car park overflowed and still they came. Now by any yardstick of marketing and enterprise, the event was a spectacular success. As a means of introducing people to red kites it was less satisfying. No-one saw a wild red kite. They saw only red kites with plastic wing tags, birds that have been handled at the nest by people, fitted with lumps of plastic that, however much science may protest, can only impair the birds in flight (wings are designed to operate with feathers and wing tags are palpably not feathers) and *must* from time to time result in death or injury by snagging and effectively snaring the bird by its wing.

By the time the tenth red kite had perched in the trees near the feeding station, it had begun to trouble me that the organisers were encouraging the birdwatchers to look for the coloured wing tags among the branches as a way of spotting the bird. But you're not spotting a bird. You're spotting a piece of plastic. It has nothing to do with the wild bird. In this part of Perthshire, the wild bird does not exist, because, as one of the organisers said in response to a question about how many kites were wing-tagged, "Oh, all of them."

What every single bird?

Yes.

Every single bird in every single eyrie?

Yes.

That is the official policy of the red kite reintroduction programme. And it will continue to be until there are simply too many red kites to monitor. Oh happy day. Then and only then will wildness begin to have a look in. Then and only then can anyone claim that the red kite has been successfully reintroduced – when it starts eluding the grasp of the reintroducers.

After two-and-a-half hours in the hide, during which time I had not only grown uncomfortably cold but also just plain uncomfortable at the endlessly repeated gospel of the wing-tagging fraternity, I got up and left the hide. No kite had come anywhere near the feeding station in that time, although we were repeatedly assured they would come in any minute. And maybe they did, but as soon as I closed the door on the hide I had my doubts, for there must have been two hundred people standing *outside* the hide and in full view of the kites in the trees. At least half wore purple or red or yellow or white jackets, and they were many more times more conspicuous at a distance than the plastic tags on a red kite's wing. Children bounced eagerly around the throng, and the general hubbub must have been audible for half a mile. Oh, people saw red kites all right, at a distance, in the trees, and maybe that is as close as they will get to a red kite in their lives. But if anyone had spoken to them quietly and one at a time about patience and how a little learning and the absence of crowds can bring you very close to red kites indeed, the needs of both wild bird and birdwatcher would have been infinitely better served. I know a farmer near the osprey wood who has known red kites follow his tractor when he ploughs. I know a keeper in the hills near Balquhidder who watches red kites home in on the sound of a deerstalker's gunshot to be at the head of the queue for the discarded gralloch. And the red kite used to scavenge our mediaeval city streets, a relationship with people commemorated in the gold-painted sculpture of the bird on a tenement wall of Edinburgh's Royal Mile. The red kite has always been comfortable around people when both are simply going about their business. But this new phenomenon of conservation to turn the red kite into a source of mass entertainment is a profoundly uncomfortable lurch away from anything that could be described as a thoughtful relationship with nature.

A second group of birdwatchers a little distance away from the hide, and equally unscreened, clustered around another organiser who had set up not just telescopes but also a radio-tracking device, for some of the birds are also fitted with transmitters, so that their movements can be tracked at anything up to 30 kilometres. I shrug my bewilderment at all this tampering with a wild creature, all this determined dismantling of its mystery. It has the distasteful air of a game employing expensive toys, the object of which is to render the wild tame and the free tethered.

As I drove home I made a promise to myself and to all nature. Never again would I report sightings and discoveries of birds and eyries to members of the conservation establishment, not if this is the use to which such information is put. There is a kind of betrayal at work and I find it loathsome.

I turned aside at the head of the loch. I felt bruised by the afternoon and in need of healing. I turned aside for the one place in all the Highland Edge where I know I can find it. It is as reliable as fresh air and rain and it did not disappoint me.

The place is as close to me as my shadow. There is no weather, no season, no hour of day or night that I am unfamiliar with along its boggy, tussocky, tree-lined shores and reedy shallows. Now, in the half light of a cold and clouded January dusk, the lochan is iced-over and gray, paler gray at the far shore where the ice is thicker, darker gray in the middle but weirdly defaced by hundreds of parallel scratches the same pale gray shade as the far shore. As the ice forms and thickens it also vocalises. There is something of the madness of jackdaws in its voice. It ricochets and echoes, the sound of it wheeling away from me across the surface where the wind shreds and scatters it.

Behind and ahead of me, well-treed hills rise steeply and darken into a steely sky. Two ravens, one fifty yards ahead of the other, cross the glen conversing across the airspace. A heron crosses the lochan at the nearest thing to zero feet it can get away with, wingtips almost brushing the ice, screeching disapproval. Ice is not in the interest of herons.

These are all that move through the dusk, these and my own dark and purposeful shape in the edge of the trees. My purpose is just out there, on the lochan, but first I want to reach a wispy screen of birch and alder about twenty yards beyond the edge of the forest, more a fragment of the lochan's landscape than the forest's. The birches struggle in ground this wet, the alders thrive. Even in their spindly winter clothes, they serve to break up my shape while I cross the difficult open ground.

It is a slow, careful twenty yards, placing each boot gently, trying to avoid the spaces between tussocks where the ice lies ready to crackle and crunch. I don't want crackling and crunching at this stage. But I fail to see the snipe. It leaps up from under my raised boot, hurtles forward, slaloming between an alder and a birch at grasstop level, swerving left at the edge of the lochan, 50 flat-out yards north then thumping down into the grass again. My mind suggests to my leg, very reasonably, that it should put my raised boot back down on the ground.

Be still. Calm down. Start again. Five more yards, five more paces, no more snipe.

At last I can lean against the trunk of the sturdiest alder and settle. This is my idea of a hide.

Fifty yards ahead, out on the ice, seven whooper swans sit on a patch of water they have kept open simply by constantly disturbing it. They sit with their necks laid back along their spines, their heads pillowed on sumptuous depths of feathers. I can see the eyes of three of them, and they are open. Occasionally a neck half rises, a head appears, and the yellow blaze of its wedge-shaped bill is vivid as moonlight, brighter than a kite's wing tag. These yellow blazes also impart information (they are all different shapes) but only from swan to swan, which is their purpose and their mystery. But the painter-naturalist Peter Scott made reference portraits of whooper swans at Slimbridge, and could identify many individual birds that way, a happier solution than wing tags. The neck subsides, the head vanishes, the yellow blaze is smoored. The open water is jet black, the swans startling white against it, the necks mostly grayer but two of them are stained pale orange from summer months feeding in the iron-rich waters of their Icelandic homeland. The staining fades in the nomadic winter months. Each bunched swan shape is mirrored in the water, and as you watch you realise that they rotate very, very slowly in response to the one foot that works to the gentlest rhythm underwater, the absolute minimum effort required to keep a patch of water a few inches deep beneath them.

Half an hour slips past, an eternity in gray, imperceptibly darkening. Then something changes.

A head rises on a half-unfurled neck, asks a monosyllabic question – "Whoop?" – like a soft clarinet. The yellow blaze glows. The neck straightens, becomes surprisingly tall. The bird rises on its tail, opens and flexes its wings – they crack like sheets in a wind – then subsides back down onto the water. The other birds stir and look around, scattering a few more monosyllables among the sudden ripples. The first bird to call now steps clumsily up out of the water and onto the ice. Its voice grows querulous. Agitation ripples through the flotilla. One by one they follow their leader onto the ice and stand.

What just happened?

Have the swans detected some flaw in my own stillness?

Have they known I've been here all the time and their tolerance has worn through? Have they been troubled by some other presence I have failed to detect?

Is there a fox out on the ice?

An otter?

Or have they simply decided that their position is too difficult? Perhaps their sophisticated knowledge of weather systems informs them the night will intensify the ice beyond their capacity to keep it at bay?

Or did they never intend to spend the night here anyway? They keep this information to themselves.

Out on the ice, walking warily, slipping often, calling brassily now (a hint of flugelhorn as I fancy it), the swans begin to demonstrate the emergence of a collective will. I can almost see it at work; I can certainly hear it. The afternoon grows dark and very cold. The ice sings as it thickens, and at each note one or more of the swans responds. The world reduces to three basic simplicities: the swans, the ice, and the dark edge of the trees where I stand. It occurs to me again how much I like being at the edge of things, the no-man's-lands where I can step between elements or landscapes or worlds and feel the connections between them, feel nature swirl seamlessly among them all, pinning the ice sheet to the bog and the bog to the wood, pinning the Highland to the Lowland, the land to the sea, the island to the archipelago.

The swan voices grow to a continuous clamour, necks stretch and lower, feet pound the ice and wings cleave the air. The sound of it all is my favourite anthem of wild places, twice the anthem as the ice picks it up and echoes it. They are airborne in fifty treacherous yards, heading north into the wind. They make a wide arc and fly south for the open waters of the big loch. They hurdle the lochside trees then drop to wavetop height where they fly more efficiently, then they land on the open water. Their voices drift up the freezing air…

…Late at night with a dram in one hand, a pen in the other and a pad of paper on my lap, I rewound the day in my head. I set the crowded hours in the red kite hide alongside the solitary hours with the swans. Several hundred people will have gone away from the red kite open day happy that they saw kites and kept the cheerful company of others who were happy to see kites. It will not have mattered to them that the kites wore plastic badges. Some at least will be pleased that they were able to use the plastic to find the birds in the trees. I considered my own discomfort in the face of all that, and find it rooted in the unbridgeable distance that such an occasion puts between nature and me. There is nothing more precious to me than the removal of that distance. The art of *becoming* nature is what I practise, and to do that you have to be prepared to keep nature's company on nature's terms. The red kite open day achieved precisely the opposite and was quite without value to me. By comparison, the time I spent with the swans was rich and vivid and precious. It lingered long into the night and I awoke wondering where and how they spent the night. So I went back at first light, but the place was swanless. The ice had thickened in the night into small ripples. But where the swans had kept the water open there was a perfect circle of crumpled ice. Within the circle lay a smooth sheet of darker, thinner ice.

If I had not spent the dusk with the swans, I am not at all sure I would have detected the circle, or if I had, I am not at all sure I would have read its meaning correctly. But now I have added to my store of knowledge about this place at the heart of my territory. Now I know how a watersheet like this freezes variously when swans work on it, and how the swans' ice-free circle looks in the morning after, when it's swan-free.

8
The Hill of the Ravens

A RAVEN FLEW HIGH across the glen making for Beinn an t-Sithein. It let fall its two-syllable contralto calling card as it flew. *Kruuk, kruuk.* I called back, aiming my imitation at the required pitch. The raven's characteristic croak is one of the easier bird calls to imitate, and the raven is perhaps the most likely of all birds to respond to a non-raven trying to sound like a raven. But the raven also has a sense of humour so I am never sure whether it is inclined to respond to me because it has rumbled me and enjoys the game or whether I have briefly fooled it. It also has a formidable vocabulary, and the flying croak is only the most familiar of its sounds to non-raven ears. Studies around the Canada-Alaska border suggest that it has the largest vocabulary of all non-human creatures. I wouldn't know, but I've heard some of the evidence and I am impressed.

So a raven was heading straight for Beinn an t-Sithein and at the sound of my fake croak it looked down and diverted briefly towards the source of the sound, flipping over and diving down and changing the pitch of its voice to something higher and softer as it dived. It levelled out a couple of hundred feet above me, and as I called again it eyed me silently, climbed and readjusted its course for Beinn an t-Sithein where I lost it against the trees. I decided to go and look for it.

The name means the Hill of the Fairies. God knows why, for all its landscape's other features were named after natural characteristics or creatures. Beinn an t-Sithein's slopes include a Breezy Hill, a Crag of the

Torrent, an Eagle Hill and a Raven Crag, and a neighbouring hill has its Wolf Crag, but all these defer to the Hill of the Fairies in terms of their landscape significance, and there is no-one left to explain why. Nor for that matter why variations on the fairy theme occur throughout the named landscapes of Gaeldom, up to and including Schiehallion.

But I am inclined to think of Beinn an t-Sithein as the Hill of the Ravens, and I named it that way in my head for the birds that used to nest in a heroic little tree on the mountain's summit profile. You could not climb the mountain in the nesting season without one or other of the adults – sometimes both – performing athletic outrage in the airspace above your head. Then some old storm or other, or perhaps just the passage of time, rearranged the tree's limbs including the crucial support for the nest, and eventually it fell from the mountain's profile. The ravens are still around, but now the nest site is more discreet, and one that eluded me for long enough. Each time I scanned the blunt profile of its small tilted plateau just below the summit, I would still expect to see ravens. If there were none the mountain looked incomplete. And then one day it occurred to me that without the knowledge of where the ravens nest on the mountain – if they *were* still on the mountain – my knowledge of the mountain was also incomplete.

I do not use the word "mountain" lightly. Some hills are just hills and some are small mountains. At 1871 feet Beinn an-t-Sithein may be little more than half the height of a Munro, but it has the air of a small mountain and puts many a Munro to shame. It stands apart and pores over my territory in every direction. From below, your eye tells you that the conspicuous south-tilting rock upthrust on the skyline is the summit, but the tiny cairn at the far northern end of the broad ridge is higher, higher but unsung.

I scan the mountain often. It dominates the landscape of the swan loch. This means I have laboured in its shadow and its presence for thousands of hours these last thirty years. When the swans' nest bay is quiet, I often turn my attention to the mountaintop for a change of scene, a change of light, a change of direction, a change of preoccupations, black birds rather than white. I am inclined to wave to it, a gesture of fealty. Besides, ravens have a pedigree as associates of the gods. Thor had two of them that he sent out into the world to gather information, and when they returned they whispered in his ears all that they had found. If you doubt that ravens can whisper, you have never sat alone and hidden and silent near a mountain roost and eavesdropped on the small talk of their most intimate moments. Such legends were the creation of story-tellers but they grew out of the known world. Their authors were the first nature writers. Some of today's finest nature writers are story-tellers too, and poets, and their seedbed is what they know of the natural world.

I stop short of the ancients' notion of gods inhabiting mountain summits (although anyone who works closely with nature in a mountainous landscape cannot fail to have some sympathy with the idea), but for a long time now I have accepted landscape as a spiritual force in my life, both comforting and inspirational. So those elements of a particular landscape that define its character hold particular significance for me and I treat them deferentially. It follows that within my own working territory, their significance is further enhanced, and my Hill of the Ravens is the centrepiece of all of it, a fulcrum of all that land from the badger wood in the south to the eagle glens in the north. These are mostly jostling hills linked by broad ridges and divided by deep glens. But Beinn an-t-Sithein stands apart. To climb it and look around you is to begin to understand the nature of the land that is the Highland Edge and how it came into being.

For two years it was framed by a window in the house I rented, and establishing its mood was the day's first task. My mind's eye would place me on the summit ridge and I would look at my distant self through the vapours rising from the day's first coffee. Now that I have flitted half a mile, I have to walk fifty yards from my doorstep for a good view of the whole ridge, and my window frames Stob Binnein instead, which is 2000feet higher and a mountain cut from a very different cloth.

Climbing Beinn an-t-Sithein is usually a leisurely expedition. You can be up and down comfortably in a spring or summer evening and I have often done just that, but it always takes longer than I expect it to, like all time spent in the company of good friends. There have been times when I have lingered to watch the moon climb among the mountains oblivious to the gathering darkness behind my back, then been grateful for its light as I climbed down. But May is the mountain's shining hour, and it was May the day I decided to go and look for the ravens…

…Much of the mountain is forested. I like it best where a track begins to climb among tall, well-spaced, spring-green larches, new grass thickening the clearings, primroses and wild hyacinths catching shards of sunlight where a well-filled burn threads silver and gold into the green weave of the mountain's skirts. A red squirrel galavants among the high branches on preposterously clawed toes, discards half-eaten cones to the track 50 feet below, sprints flat out down a dead-straight trunk then comes to an impossible stop six feet from the ground and bends its head up and away from the tree to eyeball my sudden presence in the intimate map of the treescape it carries in its head.

It is close enough to fill the binoculars. Its eyes are pale black and dark black. Its upcurved throat is brilliant white. Its ears carry the pointed tufts of a lynx. Its tail is flat against the trunk and points straight up, back the way it has come. Its limbs are splayed wide. It contrives to be dead still and desperately restless in the same instant. Its coat is glossy in the early morning sun. It looks like a cartoon of itself, skinned and hung up to dry on the tree by its tail.

We are perhaps ten feet apart. I stand slightly downhill from the base of the tree where the squirrel has stopped, precisely at my eye level. It stares, unblinking. Where my face should have eyes it has binocular lenses. These seem to fascinate it and I wonder if they are catching the light and bouncing it back. Seconds pass. Nothing moves. I hear a woodpecker nearby, the wheezy screech of a jay, a chiff-chaff, a cuckoo. But they reach my ears as far off sounds divorced from this contest of stillnesses and wills. I have all day. But the squirrel is better equipped to be a statue.

An absurd memory surfaced while I watched. I was in Alaska, talking to a woman who was teaching the almost extinct craft of sewing animal skins to school children. She showed me a ceremonial cape used in traditional dances made from forty-eight squirrel skins. Then she handed me a single untreated squirrel skin.

"It doesn't look much like a squirrel skin to me," I said.

"That's because it's inside out," she said and laughed delightedly.

"So you need forty-eight of these for one cape?"

"Oh no, I need sixty-four, so that I can be sure I have forty-eight that will match."

And she laughed again.

And I look through the glasses at the red squirrel on the larch tree on the lower slopes of Beinn an t-Sithein, and it looks more like a cartoon of itself than ever, more like its own skin hung out to dry. I wonder what it would look like inside out or sewn with forty-seven of its matching kin into a ceremonial cape for dancing. Then the squirrel twitches.

At the end of his twitch it has turned through one hundred and eighty vertical degrees and now faces up the tree. I swear I could not discern a single individual movement. I would love to see that manoeuvre again on film in slow motion, see the order of the limbs as they effected the change in position, for the speed of it confounded any possibility of thoughtful observation.

The squirrel is a splayed statue again, but no longer looks at me. Instead it stares across the track to my left and uphill. I inch the binoculars down from my eyes and follow its gaze. A roe doe has stepped from the nearest clearing onto the track and has suddenly interrupted her own preoccupations to confront the unnatural stillness of the two creatures before her. Perhaps the squirrel twitch alerted her. So now the stillness is threefold, for the doe has frozen with one forefoot a few inches ahead of the other. Her nose is working and her ears are wide and forward. By chance she stands in a spotlight of sun in her deep red-brown summer clothes, a walking encapsulation of the forest's spring spirit, the dowdy gray of winter a discarded memory.

Then the thing dissolves. The squirrel turns away into the shadowed side of the tree and is next seen leisurely climbing another larch 20 yards away. The roe steps silently back into the clearing, pauses between trees to look back at me along her spine, then soft-shoes calmly away. I follow her with my eyes then turn to look for the squirrel in time to see a vandalised larch cone tumbling through the branches of the next tree but one. I look up the empty track and the sun pours tilted pillars among the larches and the shut-out sounds of birds rush back into my ears. I walk onwards and upwards smiling.

The last of the forest is the worst. The path dwindles suddenly to the width of a couple of pairs of boots, the trees thicken and darken and drain themselves of every colour. The climber enters a wooden tunnel, haunt of a gloomy silence which the groaning of broken trees only serves to deepen. I have walked here in a blizzard when the snow penetrates as thinly as the falling feathers of a preening swallow. The place is a grim nothing, a long zero slung beneath a canopy of cramming trees. Nothing moves here. Nothing sings or crawls or creeps or walks, only those humans who are urged this way by a Forestry Commission path. It helps me to know that at the end of it lie the familiar glories of the mountain's upper slopes.

Sunlight and birdsong conspire at the far end of the tunnel. Stepping out of the deadly gloom onto the warm, airy, living, song-drenched hillside is a kind of re-birth, a sublime transference from a place of no colour to a sudden land of Wordsworthian yellow. Not daffodils, but I turn into the hillside and a little den beside the path is patched with thousands of primroses. Suddenly Wordsworth's over-celebrated, over-quoted wandering-lonely-as-a-cloud poem that had been thrust down my throat as a child (and which I have loathed ever since) thrusts itself into my head. I rotate the lines the way the way they were inflicted on me fifty years before by a teacher who seemed to believe that poetry was an exercise in linguistic regurgitation. Suddenly I come up against its two-line hidden treasure:

They flash upon that inward eye
Which is the bliss of solitude;

That, in the estimation of the poet, is the true worth of his encounter with the daffodils, the retrospective appreciation of the moment. Suddenly, after all these years of antipathy, Wordsworth and I have found common ground. For if you keep nature's company with all your senses alert, its revealed secrecies will sustain you again and again and again. You build a reference library in your mind and draw on its entries more or less at will with your inward eye, and these enrich with the passage of time. You remember not just the moment of revelation but how you worked for it, then how you reached for the right words to try and make it immortal. And if like Wordsworth your words are on the lips of half the English-speaking world, you might dare to hope that somehow they will elude forever the notice of the kind of teacher who thinks that poetry is an exercise in linguistic regurgitation.

So I step among the primroses when all at once I saw a second crowd, a host, of peacock butterflies fluttering and dancing in the breeze, and they are every bit as mesmerised by the primroses as I am myself. I decide to try and photograph a peacock on a primrose. My technique is simple. (All my photograph techniques are simple, my preferred means of attempting to thwart technical incompetence.) I choose a well-lit group of flowers, tidy away a few pieces of stray grass, sit down beside it and wait for a peacock to alight on it. It takes twenty minutes. In the end, the photographs were less than satisfying, but the twenty minutes of sitting still were priceless. I watched the butterflies, noted the islands of wood anemones and violets among the primrose sea, the tiny cluster of alpine lady's mantle leaves by the burn, the songs of three different warblers, the power climbing of a pair of buzzards, a stag in velvet and languid in the sun. Finally the calls of two unseen ravens – one a tone-and-a-half higher than the other – remind me why I had set out that morning. The transformation from the blacked-out tunnel of trees is now complete.

The top of the mountain's signature rock outcrop is matted with alpine lady's mantle. When the wind blows the mantle over the lady's head, she reveals silver petticoats, and these are her modest glories. The flowers when they emerge in summer are merely modest. The lady has magical powers. Her virtues used to be boiled into a drinkable solution that healed wounds and stemmed internal bleeding. And in eras when sword wounds were a fact of life in Highland Scotland, bandages were dipped in the solution before they were applied. Unwounded and not bleeding internally, it pleases me greatly just to keep her company in the high, wild places of the Highland Edge. Out along the ridge the little white and yellow flowers of cloudberry are just beginning to bloom, and these too are symbolic of the kind of landscapes where my preferences lie. The first emergence of both those plants high on the ridge of the mountain mark small turning points in the wild year. Pipits and skylarks voice their own commitment to the burgeoning spring, there is warmth on the wind, and a buzzard is using it to climb leisurely thermals up into golden eagle airspaces. But where are the ravens?

I put a raven call on the air. Sometimes if they are within earshot but out of sight, they will come to see who calls and what their business might be, especially if the caller sounds as if it might infringe the nesting territory, always assuming there is a nesting territory on the mountain. I begin looking as far away from the most popular parts of the mountain top as I can get, looking for likely cliffs, and also bearing in mind that the present raven generation might well have inherited a penchant for trees from their forebears. And as I looked I called, hoping to elicit a response, a clue, a giveaway. It is haphazard and unscientific, but then I am an unscientist, every bit as unscientist as a raven, and the nature writer's cause is better served by adhering to a kind of wild improvisation than to principles and routines of science. Besides, I am here to *write*, to try and persuade the mountain to share her secrets with me and to write with them. We are each other's raw material. She makes what she can of me and I make what I can of what she reveals to me. Can I tune in to her softest speech, catch snatches of her song? I am here to be poetic if I can, not scientific. The mountain is the hub of a nature poet's territory, not a biologist's, nor even a naturalist's. The nature writer is not a naturalist, although he will hopefully acquire some of a naturalist's skills as he confronts and conspires with nature. But his needs are different. He does not need to put his human stamp on nature – count things, collar and ring and wing-tag things, weigh and measure things, make surveys, reach conclusions, publish reports. Instead he needs to feel things. He needs to take nature's hand, look into her eyes. Then he needs to take what he feels and sees there and make words of it on paper; to bring news of it to his own species; to show that his species is still capable of intimacy with nature for all the distance it has imposed on the relationship these last few centuries; to show that he is still nature's brother.

I look among the known cliffs and crags and lesser outcrops, especially the outpost the map calls Creag an Fhithich, which is Raven Crag, and once there must have been a reason why, and there might still be. But there is no raven nest there. No bird answers my call. I sit for a lunch hour, looking west to where the Allt Fathan Ghlinne and the Allt a'Ghlinne Dhuibh join forces to become the Calair Burn. Down there last summer, among the rocks and pebbles by the river bank, I had found a small, oval, flattish stone strikingingly patterned black and gray and white by lichens. The black formed a sky, the gray formed a landscape of river and hills, and the white…the white was a raven against the black sky. The stone sits on the windowsill above the desk where I write. I take it down often to marvel at is casual art. A naturalist looking over my shoulder sees no such thing. But he knows how lichens work:

"*Lichens are dual plants formed by a fungus and a plant of the alga group. The algal cells contain green chlorophyll and manufacture sugars and other compounds to sustain the plant. The fungus provides shelter for the algal cells and prevents them from drying out. The various shapes of lichen depend on the fungus. There are 1355 British species of which 500…blah, blah, blah…*"

There are times when too much information is not helpful, times when it blinds you to nature's mystery. I see a white raven in the palm of my hand, a gift from the Hill of the Raven.

I walk on thinking about how I would feel if a white raven suddenly appeared in the sky over Beinn an t-Sithein. Now that would be a bird of omen! The Vikings voyaged with a raven's head on their longship sails, and that emblem was fearsome enough to those on whom they imposed themselves. Imagine if there was one boat in the flotilla that bore a black sail and a white raven…

Or, imagine if the Tlingit Indians of North America (whose tradition was that it was not God who made the world but Raven) should suddenly be confronted with the sight of a white raven. Surely they would have called it sacred and regarded it reverentially, even fearfully, an emissary of the original Raven returned to pronounce judgement? Science, confronted by the same creature, would offer a lecture on the deficiencies in pigmentation of albinism.

I call again to the empty, open hilltop, but some freak of rock formation throws an echo of my "raven" back at me, so of course I call again and again and the mountain air rings with my reverberating ravens, and then in my head I hear the echo of a different raven. I was in Norway, making a radio programme about wolves, and talking to two film-makers as we walked and drove around the landmarks of the wolf pack territory they were filming. The territory, for between five and seven animals, was two hundred square miles. They used traditional tracking skills to find them and film them, they revered the wolves' mystery, they acquired their knowledge from their own five senses and from an intimate knowledge of the ground, and from the ravens.

They had begun to notice that when ravens found a dead moose they put the word out by calling. They summoned other birds to the feasting table. The wolves already knew this, of course, and if they hadn't actually killed the moose or detected its smell, they would home in on the gathering of ravens. So sometimes the ravens also led the film-makers to the wolves. One of them prompted his colleague to tell me how many ravens he had put up from a moose carcase at the Canyon of the Trolls.

"Seventy-seven," he said. "Seven, seven," he repeated carefully, to ensure I had understood his English.

The canyon is a phenomenon of rock and pine trees and ravens. They nest in the walls and when they call the echoes ricochet off the rock faces and amplify wondrously. I tried to imagine what seventy-seven ravens would look like and sound like, imagined the distant wolves stopping in their tracks, knowing its meaning, reaching an instant consensus, changing direction, picking up the pace. And all the while the horde of ravens grew clamorous and the canyon walls reverberated with echoes and echoes of echoes.

And now the sound and the spectacle of it all revisits me as I walk along the summit ridge of Beinn an t-Sithein. With my head so full of ravens, wolves and Norway, a fast black shape suddenly appears from below, calls once, then vanishes back the way it has come, dragging me back to the landscape of the Highland Edge as it flies. I look at the possibilities, consider what I know of the unseen hillside below, and plan a detour. There *is* a place, and I know how to get a good look at it under cover if I drop into the high edge of the forest again. The detour takes ten minutes, much of it trying to walk doubled up under spruce branches that reach for your eyes, your cheeks, your hands, your hair, that snag your clothes and shorten your temper. It is not my favourite part of the mountain, nor is it a place where you can travel softly, so if the hunch that I'm backing is correct, my presence is being announced to the ravens by the spitting of broken spruce twigs. I reach the edge of the trees, check my bearings and reckon I'm about 50 yards short of the crag I'm aiming for. And my hunch is also 50 yards out, for at once two ravens speed silently from the rocks directly above me. I put the glasses on the spot and find rocks streaked and splashed white, and then the twiggy mass of the nest. If there are nestlings in it they are keeping low and silent. I retreat at once, an undignified and spruce-infested retreat that takes me well away from the nest and precludes the possibility of being seen from above by the ravens.

Back on the open hill, elation sets in. I also work out a mental map of a much less painful way of approaching the site through the trees so that if necessary I can keep an eye on the progress without imposing on the birds, although I never much cared for lingering near nests. I walk down past the primroses and re-enter the tunnel with my head full of ravens again, for I know they are secure on the mountain, and in my own mind at least that makes the mountain complete.

I walk with music in my head too, as I often do, this time a jazzy, bluesy thing that I try to order into some kind of sense, composing on the hoof. Sometimes I remember it when I get it home and formalise it with the guitar. Sometimes it vanishes like the echo of a raven on a mountain wind. Back out among the airier spaces of the big larches, a tiny patch of vivid blue snags in a corner of my eye, the first speedwell of the mountain's spring. I sit down beside it just to commemorate the moment, and a poem slides into place where the tune has been.

The Softly Blues

The softly blues of speedwell
by a shadow-and-sunlight mountain tree
are pianissimo, man,
Monk in a minor key.

Easy now,
with only height to lose,
I got those low-on-the-mountain
softly speedwell blues.

9
Wind's Work

NOVEMBER'S WIND GREW BIG in the afternoon. All morning it had flailed the glen with squalls of black rain. Mountainsides advanced and receded, blurred armies of heaped curves that faced each other across the space where the glen should be, blunt gray ghost armies, their every autumn shade withered away by the rage of wind and rain. I watched from my desk in the window that faces the glen. The window shuddered, and so did I, working with the light on at noon.

Two hours later I heard something change, or at least I heard a sudden absence. The voice of the rain was gone. But now the voice of the wind roared. I looked up and saw light in the sky, an edge of gold on a hilltop, a blue hole in the tumult of clouds. I grabbed jacket, overtrousers, wellies, binoculars, camera, and I went out. The wind had grown so big it had blown the rain from the sky.

I climbed the hillside by the forest track. At the first stand of tall trees, the wind's voice was like surf on a Hebridean shore; strange to have surf break eighty feet above your head.

The hillside is my back doorstep (the glen is the front doorstep); it's where I stretch my legs and clear my head on writing days. For a Commission forest it wears better than most. There are trees of every age, and pockets of Scots pine and oak and birch and rowan as well as spruce and larch. Brambles and raspberries and bracken and grass thicken the wild corners, orchids throng the damp places. And I have seen these: robin, wren, dunnock, blue tit, great tit, coal tit, long-tailed tit, siskin, crossbill, bullfinch, yellowhammer, chaffinch, goldfinch, greenfinch, goldcrest (once a flock of 50, a tiny horde), twite, redpoll, tree creeper, raven, crow, jackdaw, jay, great-spotted woodpecker, green woodpecker, heron, sparrowhawk, kestrel, buzzard, osprey, peregrine falcon, cuckoo, golden eagle, tawny owl, barn owl; and (in flight to somewhere else) mute swan, whooper swan, Canada geese, greylag geese and red kite; and red squirrel, red fox, red deer, roe deer, stoat, weasel, pine marten, and a mountain hare I presume was lost.

So on the afternoon of the big wind I was confident of encountering some refugees from the storm at the very least. But in an hour and a half of walking, I saw no living thing that hadn't been uprooted or wrenched from the ground or from the trees. Tumbling larch and pine cones that often indicate a red squirrel in the high branches now hurtled past my shoulder at the speed of the wind, pursued by leaves and small pieces of tree. I half expected to be passed by one with the red squirrel still attached.

At the highest open area of the forest the wind tore open the edge of a plantation. I watched it barge in to the dark depths of trees and force a parting, a Moses wind parting a Black Sea. The trees bowed to both east and west as the wind raged north, and they cringed before it. I saw one tree snap about four feet above the ground. It had been a north-leaning curve. It suddenly became a rigid diagonal, held weirdly in place by its still living kin.

The wind had grown over three days of heavy rain at the end of the autumn's first cold snap. The rain wiped the autumn's first snow from the summit of Stob Binnein. It flooded the low-lying rough grazing at the east end of the glen to form what is known hereabouts (and for obvious reasons) as Loch Occasional. When the River Balvaig decides to encourage the process by overflowing, trees wade waist deep and fences submerge. Geese and ducks and swans appear overnight to cash in on the temporary miracle. Feeding on grass through shallow water is what they love best.

But on the afternoon of the big wind, no bird stirred, and no beast cared to grapple with the banshee. There was only the land and the wind and the rippling waters of the part-time watersheet far below. I walked into it all with the sense of being a spectator in an empty stadium for a contest of giant forces. Element versus element. The floodlighting system was faulty. There was flood. There was lighting. But the lighting came and went across the sky and darkness rushed about the land in mad pockets a square-mile at a time. Everything that I could see was on the move, but nothing was changing position. Cloud rose and fell on mountainsides. Trees bent and straightened and bent again and their summits whipped across the sky, dancing to the wind's grotesque tune. Tiny waves raced across the water, but the level neither rose nor fell. Hills vanished, but reappeared where they had stood before.

The day began to feel deliciously dangerous. If nothing else in the forest was out in the open and walking about, was it a good idea for me to be there? Yet I thrilled to the excess energy that billowed along the hillside and thrashed at the trees the way a rutting stag thrashes antlers at long grass. I felt supercharged by it, powered between the shoulder blades; I walked and walked, higher and higher, and the wind proclaimed itself louder and louder. It was as if the overabundance of primitive force touched everything it encountered and powered it to its limits.

A walk in that forest is normally slow and constantly on the lookout for bird and beast and flower and whatever else crosses my path. Once I was brought to a halt baffled by a rhythmic clicking sound that seemed to come from across the drainage ditch by the forest road where I walked, a single click every four or five seconds, a slight sound but so rhythmic as to be conspicuous. I tried to pinpoint its source in the undergrowth or the low branches, then a flicker of movement caught my eye beneath my feet. There was a small pool of rainwater in the ditch. A flotilla of water boatmen was sculling across its surface. Every few seconds, a single dragonfly with a four-inch wingspan made a low, lumbering approach like a Lancaster on a bombing run and snatched a boatman from the surface. The click was the sound of the attack. Seconds later it had completed a circle and was coming back for more. I started to count. Thirty! Plus however many there had been before I started to count, but certainly double that number.

Once I met a family of wrens so newly out of the nest that they could barely fly more than a yard at a time, so they stayed close to the tangle of brushwood and brambles that had been their nursery, and not more than six feet from the edge of the forest road. I stopped to watch, and the nearest fledgling was about a yard away. Its berserk parent came and stood between us and raged at me. I bowed politely, held up my hands in a gesture of acquiescent goodwill that was completely ignored (I may think I sent a warm wind to a grizzly bear across a quarter of a mile of open Alaskan water but a Perthshire wren at 30 inches was quite out of reach). So I watched with the glasses from screening bracken ten yards away as both parents crammed tiny meals into thinly clamorous open throats. Many of the spectacles of the forest are as intimate as that.

But on the day of the big wind, there was simply nothing to see but the land itself, and that was an insubstantial spectacle. The walking became an end in itself, and when that happens I get bored. I don't like to admit to being bored in nature's company. Besides, walking has always been a means to an end for me, the only natural way to get into a position where I can come close to nature, but it is by being still that nature grows confiding and shares her secrets. Finally I found a space that looked down the glen and in the lee of a thicket of tubby little self-sown spruces. I sat, and the wind was suddenly elsewhere.

But the wind was running the show, orchestrating the day, and all nature danced to its tune. The speeding, rising, falling sky was the wind's work. The on-off light was the wind's work. The dance of the hills in and out of the murk was the wind's work. The soundtrack in the trees was the wind's work. The wavelets on the distant water were the wind's work, and so was every rush and rivulet of the loud and mobile air. Even the absence of bird and beast was the wind's work. So to sit out of the wind was to be excluded from every aspect of nature that the day had to offer. I watched a rainbow stand above the glen against a dark-blue mass of cloud the shape and size of a Himalayan peak. Then I saw it shred as the Himalayan one crumpled and the afternoon's second hole of clear sky hurtled among the clouds. Wind's work.

I grew as unsettled as the day. I came out of hiding and walked back down through the forest to the window and the desk. Halfway there, the rain returned and the south-west blackened behind it. The last half-mile was like a black and sodden fog. There was only the next fifty yards of track, the vague and colourless shapes of the nearest trees, the sound of the wind and the sound of my hurrying feet. In ten minutes I was soaked, but for better or worse, I was finally embraced into nature's mood on the afternoon of the big wind.

Back at the desk, the window still shuddered, but beyond it, where the glen curved away west among mountains and lochs and forests and skies, there was simply no view at all, nothing to see but the blurred gray veil of rain that had consumed the known world and made a river of the road. It got dark early that evening.

I woke suddenly in the wee small hours. The sky was all stars, the mountains solid and black. Behind one of them there was a stain of yellow where the moon was setting. I opened the window and let in the silence, the stillness. Then I heard an owl, a fox, some passing geese.

10
Orchid Time

THERE COMES A POINT in the too-fast unfurling of summer when I want the thing to stop. Every year, the moment those zesty days and long, dawdling evenings of late May and early June start to melt into too-full, too-lush, too-overburdened-with-green midsummer, I have the same troubling feeling that I haven't made the most of the year's great festival of life-making. Could I have spent more time with the ravens, checking osprey nests and looking for new ones, catching up with the badgers on both sides of the Highland Edge? Where are those otters I keep glimpsing on the river holed up? Could I have helped the swans on the loch? Should I have helped if I could? (Yes.) Could I have helped more with the local eagle watch? If I had, would it have made any difference? (No. They failed again, abandoning the usual eyrie at the end of an atrocious, cold, wet spring, the way eagles sometimes do. To know why we would have to ask the eagles, but we never learned that trick yet. Good.) Why didn't I spend more time in that wood, those hills, watching rivers and lochs and bogs?

The nature writer can never spend too much time just looking, and there is no time of the year more suited to just looking than the cusp of spring-into-summer before the wild world starts overdosing and bloating on green. So I want the whole thing to stop, to linger, to spend another month or two in the golden-green time of May-into-June, and then, by some natural process I haven't quite fathomed yet, evolve seamlessly into the yellow light of September.

There are problems of course. One is that I wouldn't have a birthday, for I was a July bairn, and it seems singularly unreasonable to deny myself my own birthright. Another is that even nature needs a break and summertime is when the living is easy.

A third problem is that I would miss orchid time, and that would be unthinkable. I am an indifferent botanist by any standards, a state of affairs that frustrates me less than it should. I have trouble telling my yellow pimpernels from my creeping Jennys, for example, and as a new flowering season comes round, my tendency to forget the name and habits of something I looked up last year and the year before and the year before bothers me, but also less than it should. Yet orchids delight me, and I willingly grant them all the patience and endeavour I deny to botanising in general. No other species of flower gets under my skin in the same way.

I cannot account for this state of affairs, but every year when the first early purples magic themselves along the forest tracks and out on the mountain ledges, or when I stumble across the first marsh and heath-spotted orchids among the hill grasses (or even some of the more rarely explored tracts of my golf course with which I alone appear to be familiar), I feel a kind of childish thrill. I understand why Shetlanders greet the spring's first Arctic terns: the miracle has returned. And because I forget about flowers for months at a time, I forget about orchids too, until that first, startling return.

Some years it is a sparse sprinkling. Others, the hill grasses are thick with them, a bejewelling so intense and exotic that it seems almost un-Highland. Some years they turn up where I have never seen them before then vanish never to be seen again. Three years ago, a hidden place on a small bank by a footpath I use a lot was suddenly and inexplicably pink, an ankle deep self-sown little throng of fragrant orchids. I passed them one sunlight-after-rain evening, and my head was turned by the scent of carnations where I knew there were none. I investigated and found the fragrant orchids. There had been none the year before, and there were none the next year nor the one after.

In a good orchid year, a certain hillside of Balquhidder Glen grows them thick as stones in a scree run. I might walk there looking for something else altogether – eagles, say – and find nothing remarkable in the underfoot grasses. A week later I might return to see how the eagles are faring and find the hillside strewn with what looks at first glance like glowing coals, the haphazard fire of thousands of orchids. Immediately my preoccupation is bent low to their urgent underfoot summons instead of scanning the skies, the mountainsides, and all the intervening middle distances the eagles inhabit.

Such is the nature of my relationship with orchids. They make nonsense of my routines. Instead of sweeping the wild world with the binoculars they thrust me down on my knees. Then I fashion a makeshift microscope by turning the binoculars the wrong way round and peering through the wide end with the eyepieces jammed up against the flower. Then the world of orchids transforms into a magic garden.

A good wild flower book helps. One of the best for my purposes is *Scottish Wild Flowers* by Michael Scott. I never got into the habit of carrying field guides with me; in fact I have something of an unhealthy contempt for them and own far fewer than most people I know. (I have a perverse preference for discovering things for myself.) So when I stumbled on the orchid hillside in the dusk after a few hours looking for the eagles, I didn't have Mr Scott's book with me as usual. I swore quietly to myself (I can never see the point of swearing loudly to myself), because on the rare occasions I want a field guide in the field, my antipathy towards them never seems a good enough reason for the fact that the book in question is in my bookshelves. What good is a field guide on a bookshelf? I also swore softly because it is inappropriate to swear loudly in the company of orchids. So I vowed to return in good light with the wee flower book, which I did, on a warm afternoon with a midge-defying breeze and sunlight firing up the flowers. These are perfect orchid-watching conditions.

So as summer nudges towards July, you might find me on such a day on my knees on the hillside (the grass still sodden from the night's rain), the wild flower book, a notebook, and the binoculars in my hand the wrong way round. I have entered a world of minuscule measurements and bewildering detail, and with a vocabulary of its own.

The world knows 7500 species of orchid, Britain knows fifty-something, Scotland a dozen or so not including sub-species and hybrids designed to confuse. I have to learn to worry about leaves – blotched or not blotched (and what shape are the blotches?), strap-shaped or folded like the keel of a boat, a basal rosette of leaves or leaves that climb and sheath the stem? And as for the cluster of flowers – the spike – is it a cylinder or a pyramid? Are the flowers hooded and if so are they strongly hooded or lightly hooded, how many lobes on the lip, how long are the spurs, do they look like a butterfly or a star or a diamond or an angel?

And then there is the smell. Fragrant orchids are unmistakable, a slim pink cylinder of flowers (except that occasionally it has pyramid tendencies and you quickly learn never to use the word 'never' in the matter of orchid identification) that puts carnations in your nose. Mr Scott says the early purple orchid is "a loose spike of purple flowers, often smelling of tomcats".

"Why 'often'?" I ask the book aloud.

"And what else do they smell of when they are not smelling of tomcats?

"And how badly do they smell of tomcats?"

So far I haven't dared to put it to the test. I prefer carnations. Besides, nothing as sumptuously beautiful as an early purple orchid should be allowed to smell like tomcats. What was nature thinking about?

The fragrant orchids are everything the purple orchids are not. They stand out in a crowd of heath-spotted and northern marsh orchids, only because they are so different and there are far fewer of them. The colour is soft, pale and subtle, the scent is a wonder, and then (and here is the worth of a good flower book and the upside down binocular trick) there are the scales. Just when I thought I had gathered all the information I needed from the upside down binoculars and made all the notes I needed to make, I found that Mr Scott had added a kind of PS to his entry under fragrant orchids that referred to their "cylindrical spike in summer of rosy-pink to whitish flowers which are covered in glistening scales".

Covered in *what*?

By now I had examined three different spikes of fragrant orchid flowerheads through the wrong end of the glasses and while I had noticed that the flowers are strongly hooded and their bottom lips divide into three rounded lobes and the spurs are slender and down-curved, there were clearly no 'glistening scales'. I was annoyed. If they were there I would know. I was immediately scornful of Mr Scott, of all field guides. I swore softly again. And then I swore softly again because Mr Scott was right and I was wrong. It proved to be a matter of thinking small enough, much smaller than I am accustomed to thinking. I had not refined my gaze's focal length down enough. I had after all been looking at eagles half a mile away. Then suddenly I had to adjust to flowers at my feet, then to flowers made huge by the upside down binoculars, checking for hoods and lips and lobes and spurs. A lip is maybe half a centimetre long. And still I was not looking close enough. Suddenly I was supposed to be looking for scales on the surface of the lip. It was preposterous.

Then reason kicked in. Why would Mr Scott mention scales if there were none? What could he possibly gain other than ridicule? The most valuable single thing I have ever learned about nature watching and nature writing is to be willing to look again. Go again and again to the place you know well and look again and again and again. *"To find new things, take the path you took yesterday."* It is a priceless philosophy. You don't worry about finding your way, and the more familiar the landscape, the more inclined it is to yield secrets.

So I look again. I take the spike gently in my fingers and turn it slowly into the sun, I remove my own shadow, I tilt the flower patiently, millimetres at a time, and then, to my soft-voiced astonishment, the thing glitters. It doesn't glisten, Mr Scott, it glitters. These are *glittering* scales. They are too tiny to separate even in upside down binoculars, but they glitter like a starry night. It is simply one of the most astounding things I have ever seen. The aurora does not outshine it.

But what is it for? Mr Scott is silent. What point in the evolutionary process did nature reach that it suddenly decided fragrant orchids must be sprinkled with glitter *as well* as smothered in fragrance? Mr Scott is silent.

I decided it was churlish to complain. He did tell me about the scales after all. He made me look again. So I could either spend £50 on a specialist orchid book or I could theorise wildly for nothing. I decided to theorise wildly. It has to be something to do with pollination. Because the fragrant orchid cannot compete numerically with the hordes of commoner species whose company it routinely keeps (most of the time, although that colony on the bank by the footpath was an exclusive one), it must try harder to attract pollinating insects. The scent is part of the attraction, and may lure the insect close, but the flower is still hopelessly outnumbered. So it works a brilliant little visual trick for those with eyes to see, those insect eyes that don't need inverted binoculars because their focal lengths are already attuned to millimetres in single figures, and fractions of millimetres if need be. And the glittering scales stimulate the seeing eyes and lure the pollinators. That is the theory. It is, of course, desperate guesswork.

But there is another single consequence of the fragrant orchid's scales of which I am certain. I am, as I have said, an indifferent and forgetful botanist. But because of the discovery of the scales, the characteristics of the fragrant orchid are carved in my mind in tablets of stone. And I will remember the fragrant orchid forever.

After the orchids, the bluebells. Sassunach friends insist on calling them harebells, which is one of the justifications I use to question the sanity of some of my Sassunach friends. And the thing they call a bluebell is what lies in a ground mist in the badger wood in early spring, and what I was taught to call wild hyacinth. Botany is on my side, and so is Mr Scott. In his wee book wild hyacinths are listed in the index under 'Hyacinth, wild' and bluebells under 'Bluebell, Scottish'. This is wonderful, my personal bluebell. And thanks to Mr Scott I also know now that the bluebell is part of the bellflower family, and therefore a true bluebell, whereas what Sassunachs call bluebell is actually a member of the lily family, and no more bluebell than buttercup. Science calls the bluebell *Campanula rotundifolia*, the bellflower with the round leaves (the root leaves, that is, not the stem leaves which are like flimsy scraps of grass); and science calls the wild hyacinth *Hyacinthoides non-scripta* and I rest my case, blissfully content with the endorsement of the man who pointed me to the glittering scales of the fragrant orchid. So if you want to be botanically correct, there is no such thing as a bluebell wood. There is only a wild hyacinth wood. And there is no such thing as a harebell. Hares have nothing to do with it. But the creeping stain of the Sassunach tongue into many of the cultures of the world infiltrates Scottishness too, and compounded by an undeniable laziness among the natives, it has permitted the English names to become common currency here. It is a thing to be resisted. The Scottish names are old, they are acknowledged in almost every wildflower book you will ever pick up, and in this case they also improve on the English, not least because they are right and the English ones are wrong. So join me in championing the cause of the wild hyacinth and the bluebell, and never let the word harebell escape from your lips. Or if you really want to confuse a Sassunach on the matter, really put him in his place, quote the Gaelic names at him. The wild hyacinth is *brog na cuthaig*, the cuckoo's shoe, and the bluebell is *currac-cuthaige*, cuckoo's cap, and if the Sassunach turns on you then and demands to know what cuckoos have to do with it given your determined and principled stance on hares, you're on your own. I have no idea.

The River Balvaig is a water snake, deep and dark amid well-treed banks, and willing midsummer host to bluebells. Even as the orchids wilt and wither, the bluebells materialise in small clusters and clouds of restless bellflowers, blue campaniles of flowers to ring in the green fulfilment of high summer. On the top of the riverbank they are sparse and ankle-high, but on the sun-facing slope they dance waist-high, hundreds of bells astir in a tight bush of blue to make you gasp and smile. The winter river floods spectacularly and quickly here, but in summer the river has a lazy *Wind in the Willows* feel and curlews and lapwings nurse new broods on the flood plain, which becomes a sweet meadow of meadowsweet. On a hot late July late afternoon, I was out among the bluebells and the riverbank alders looking for otters, or at least for signs of them (for these are not the most promising of otter-watching conditions), and for some indication that the neighbourhood kingfishers had prospered that summer. For two hours I walked the bank, pausing often to scour the overhanging branches for kingfishers, or to read the tracks of otters in the mud, nodding aggreably to every bluebell gathering, practising my Latin – "*Campanula rotundifolia*" – I said, and "*currac-cuthaige*" too, for my new knowledge was like a persistent itch demanding to be scratched, and Mr Scott's endorsement of my lifelong denouncement of the errant Sassunach was a warm glow within.

Slowly the warm tranquillity of the hour imposed itself, and my steps dawdled and then I simply sat in the shade of a towering ash, solitary on a bend in the river, a landmark of a tree these last hundred years for every creature that passes. Thick shoals of tiny fish cramped as sardines in a can, but thousands of fish in each shoal, flirted in and out of the shadow and the sunlight, an impossible coherence about their brief and flat-out mass movements. What were they doing? Hunting? Enjoying the play of the light? Fearing the presence of my shape on the bank? Was there a pike in the depths or an otter lying up in the roots? Did they see the sunlight ignite the bluebell dance a few feet above the water, and did they dance to its silent tintinabulum?

At times like this, I can almost believe in high summer. But it needs the sun and it needs an easy breeze to thwart the midges, it needs honeysuckle and meadowsweet and blaeberries on the hill and raspberries by the forest track, and it needs an idling river and swerving shoals of wee fish, and the prospect of otters and kingfishers. And it needs bluebells in clusters and clouds.

For perhaps another hour, I sat on the riverbank enjoying the sun on my back or the shade of the tree, making free with both like the shoals of fish. And mostly I was absorbed in writing in the notebook on my knee, but occasionally I looked up to watch a tall and dense blue cluster of bluebells on the far bank that glowed a deeper blue than its neighbours. It stood near the top of the bank and above a deep dark pool, and the sun was full on it and it swayed in the breeze and a hundred flower heads nodded emphatically. All this was mirrored and stretched far out across the river by the reflection in the pool and the sloth-slow current and the occasional touch of the breeze that was no more than enough to ripple and wrinkle the surface against the direction of the current. I was wondering idly how a good painter would begin to paint that, when the reflection suddenly broke apart in the mirror and burst into blue shards.

What the…?

Inexplicably, when I think about it now, I looked up to the living bluebell cluster on the bank to see what had happened, and nothing at all had happened there. It blazed its sun-smitten deep blue and swayed and nodded as before. But back in the pool, a small mayhem had broken out, a mayhem inflicted by the gentle passage across the pool's surface of the blunt head of a small otter. I heard my own intake of breath. The otter moved soundlessly downstream a dozen yards then dived to negotiate the spiky branches of a fallen tree. It did not reappear. I waited another hour, an intense frenzy of stillness and silence, for its return, for a glimpse of a parent or a sibling, but nothing showed. The bluebells recomposed their reflection in the pool, but the sun found a small cloud in its portion of the sky and the river dancers dimmed.

All over the Highland Edge, at every hour of every day and night, moments like that unfold as the natural world goes about its business, and mostly, because we work to a different schedule and dance to a different tune, we are oblivious to them. If, however, you are willing to sit still and let nature come to you, and if your eyes are open when it comes, the chances are that sooner or later, one time out of ten, say, you will find yourself in the right place at the right time. And if it is one of the other nine times, you have to be willing to go and sit again and it is as simple as that.

11
Thoreau's Ghost

THERE ARE OTTERS on the pond too. They tell me they have been there. Sometimes they tell me they have just left. But I don't speak otter and I can't read it, for all that they leave a great deal of information lying about which is understood by other otters. What I do know is that in the three years I have been accustomed to visit the pond I have never seen one there, despite varying the times of my visits – dusks and dawns, midwinter and midsummer midnights, as well as my preferred long working lunches with my back to the big rock on the top of the bank, my face to the sun and the view of the Hill of the Ravens, and a notebook on my lap.

Occasional strangers find me there and ask:

"What are you doing?"

"Working."

This satisfies most of them and they recognise that I want to be left to my work, whatever it may be. But sometimes someone wants to know more.

"What kind of work?"

They nod meaningfully towards the pad where a thought or a poem might have trundled halfway down a page above a botched attempt to draw a gray wagtail on its perch. In return I nod meaningfully towards the placid surface of the pond where a little grebe jostles with the rush hour throng of dragonflies among the water lilies:

"Uranium mining."

That usually works.

The pond is a mile away, and used to be half that distance before I moved my desk half a mile to where it has a view of Stob Binnein. Whenever I can't think of anywhere else to go – or if I can but can't spare the time to go there – I go to the pond. Whenever I have a day so uncluttered that I can eat my lunch anywhere I have in mind, I go to the pond, because that is where I have in mind. And at odd hours of the day and night, I think of the pond and the small dramas that may be unfolding there and then, and I get up and go to the pond. And there I sit by my personal Walden, munching a sandwich or a banana or both, or cupping my hands round the coffee if it's that kind of day, waiting to see what turns up, and incubating my thoughts like Thoreau's ghost:

It is no dream of mine,
To ornament a line;
I cannot come nearer to God and Heaven
Than I live to Walden even.
I am its stony shore,
And the breeze that passes o'er;
In the hollow of my hand
Are its water and its sand,
And its deepest resort
Lies high in my thought.

I never much cared for that last rhyming couplet, but only because if you speak it in a Scottish accent (which I do and Thoreau plainly did not when he wrote it), resort has a short "o" and a chunky "r" in its second syllable and thought has neither. But quibbling with Thoreau is not a healthy trait in a nature writer, even a Scottish one, for he was one of the founding fathers of the job as we know it today, and it wins you few friends to be disrespectful to the founding fathers. Besides, I like the man and his famous book, even if he did borrow from Hazlitt his favourite line that he "...never found the companion that was so companionable as solitude".

Or as Hazlitt wrote: "... out of doors, nature is company enough for me. I am then never less alone than when alone."

And again: "...mine is that undisturbed silence of the heart which alone is perfect eloquence."

And again: "...Still I would return some time or other to this enchanted spot; but I would return to it alone. What other self could I find to share that influx of thought, of regret, and delight, the traces of which I could hardly conjure up to myself...I could stand on some tall rock, and overlook the precipice of years that separates me from what I then was."

I like to imagine that Thoreau, who moved to Walden in 1845, read Hazlitt (1778-1830) and borrowed from him, in which case that is something else Thoreau and I have in common. And he also had more than enough felicities of his own:

"A lake is the landscape's most beautiful and expressive feature. It is earth's eye; looking into which the beholder measures the depth of his own nature. The fluviatile trees next the shore are the slender eyelashes which fringe it, and the wooded hills and cliffs around are its overhanging brows." (And if fluviatile is new to you – it was to me – it means belonging to the river, so presumably his Walden pond had a river near its shore.)

I love the idea of my pond as earth's eye. It lies high on the lap of the landscape, about 600 feet up, near the wide watershed of its glen, among heaps of glacial debris and well spaced oaks and birches and a few rowans, Scots pines, beeches and ash. Mostly forested hillsides close the longer views in every direction.

The pond is not alone. It feeds a smaller, lower, more overgrown sibling, and it is fed in turn by a lochan a hundred feet higher and a tiny mountain burn. It looks as if it has been there forever, or at least since the glacier gave up the unequal struggle against the warming sun a clutch of millennia ago, but it was dug out by the estate for shooting artificially raised duck about twenty years ago. Since then the estate has changed hands, the owner doesn't care for shooting, and the ponds have been enthusiastically embraced by nature. Not just the ponds either, for the lower pond has a little outflow too and that seeps away into a wild treachery of bog and head-high reedbed. So almost by default, a little extra biodiversity has crept into the landscape of the watershed, and the otters, the little grebes, the dragonflies, the water lilies, and I are grateful. In the reedbed there are reed buntings and sedge warblers and their unskilled offspring succumb regularly to the unerring blur of the neighbourhood merlin. There are also dead-straight trails through the reeds the width of an otter. I once saw a roe deer gingering along one of these during a freeze-up, placing every foot with laborious care. Why? When she could have just walked round it the way she normally does? Following her nose, perhaps, eager to explore a corner of her territory that is normally closed to her, possibly intrigued by the fresh otter scent. She saw me when she had about ten yards of reed bed to go, and about seventy if she turned round and retreated. The trail was less wide than she was herself, and the reeds rubbed against her flanks and whispered in her ears as she stood. She froze. Then she looked to either side and saw the shifting walls of reeds. She didn't seem to fancy turning round, for that would have meant stepping off the otter trail. She didn't fancy coming straight on, for I was there, and while she couldn't smell me because the wind was in my favour, she could see what I was clearly enough, even if I was as still as the rock at my back. Then she looked back along her spine, back the way she had come, back down the otter trail, and the solution to her plight presented itself. She started to walk backwards. She reversed all the way to the edge of the reed bed watching me as she walked. She didn't hesitate once, nor did she put a foot wrong. Then she barked and vanished among the trees.

The Canada geese come in the early spring. Small skeins fly up and down the glen to the nearby lochs, but only a pair ever comes to the pond. They are incomers in Scotland, of course, outrageous imposters freighted here by transatlantic clippers 300 years ago, and much admired by the equally outrageous Victorians to ornament their ornamental idea of Highland life. Escapees from various big house collections have grown a wild population that decided not to bother migrating back to Canada or anywhere else much further than an occasional month down at the coast. They prosper here which causes purists-for-the-sake-of-purism to wrinkle their noses in disapproval.

I approve. The pair that fly up from the near south with the spring are my living link with Walden. Thoreau mentions Canada geese specifically:

"In the morning I watched the geese from the door through the mist, sailing in the middle of the pond, fifty rods off, so large and tumultuous that Walden appeared like an artificial pond for their amusement. But when I stood on the shore they at once rose up with a great flapping of wings at the signal of their commander, and when they had got into rank circled about over my head, twenty-nine of them, and then steered straight to Canada..."

This pair that come in the spring and go through every pre-nesting ritual and never nest and never stay more than a month, create precisely the same impression on a misty morning that Thoreau describes. They are imposingly large, breasting a path through the mist's eerie tendrils, scattering the little grebes, unchallenged for as long as they choose to stay, the pond at their disposal, its tiny island the perfect nest site.

But then they go, not to Canada, but down to the lochs two or three miles away, and the pond is suspiciously quiet in their absence. But while they linger, I enjoy the sight and sound of them, and when I stand on the shore, unlike Thoreau's geese, they are as likely to swim towards me as not.

On spring's first quiet, warm day, warm enough for the pond to steam, I found them in a mid-morning doze, drifting a notch above zero knots, crept close and sat, and scribbled them down in a crude sketch while they dozed. They have black eyes set in the black part of their piebald head, but when the eyes close, the eyelid is white. When they awaken, the places where their eyes are disappear, which is disconcerting if you don't know why.

I like the fleeting presence in my chosen territory of a direct link with Thoreau. For although nature writing has been around in Scotland for centuries, certainly long before Robert Burns and Duncan Ban MacIntyre took it to places it had never been before, it was Thoreau and his direct American descendants like Aldo Leopold and Barry Lopez who established the tradition I stand in now. It is unthinkable that Walden did not also colour Scottish stalwarts of the art like Seton Gordon, Fraser Darling and Gavin Maxwell who supplied the second bloodline that I willingly acknowledge in my own work, but today it is the Americans who have enshrined nature writing's place in literature, and I find myself deeply envious of the status that it carries on the American side of the Atlantic. Why Scotland should have so neglected nature writing that it routinely occupies less than a hidden shelf-and-a-half in mainstream bookshops wedged between Gardening and Pets (as though the natural world was an optional hobby) is a mystery I do not begin to understand. Our landscape is central to our idea of ourselves and some of our wildlife species – golden eagle, otter, red deer, red grouse, osprey – have a totemic significance much admired by visitors. Television programme makers and wildlife photographers regularly drool over it all, yet nature writing as literature is an idea that still escapes most Scots, most Britons for that matter. American bookshops and publishers revere their best nature writers as enthusiastically as their best novelists and poets – indeed, they are often the same people – and their universities teach the subject *as literature*. Each year produces new practitioners, young men and women who zealously champion the cause of wildness and marshal the forces of resistance to the worst excesses of their government, and in the process they produce literature. The reintroduction of wolves into Yellowstone after an absence of seventy years steamrollered opposition on a tidal wave of 150,000 letters of support from the people. Now, ten years later, the transformation in the health of the landscape wrought by the wolves is being celebrated by new literature. And all nature writing has benefited. David Gessner was a new name to me, the author of a book called *Return of the Osprey – A Season of Flight and Wonder*. According to his biography on the dust jacket, he teaches creative non-fiction at the Harvard Extension School. Thoreau must take a lot of the credit.

So when I sit a few yards away from a dozing pair of Canada geese on the shore of my local pond, and write them down and scribble their shape with an untutored hand, it's hard not to stop and think of what flowed from a situation just like this 150 years ago and a couple of thousand miles to the west.

The summer caved in – again – in August. Rain poured down mountainsides for a week, followed by two hot days, followed by Hurricane Alex when the rain hit the mountains in lumps and ripped out hillsides, started landslides that closed roads, rearranged bridges and deposited their pieces up trees and down gullies. It was rain made deafening and dangerous, and not quite what we expect in Highland Perthshire, for all that we like to think we know a bit about rain. We made the national television news.

At the first hint of a lull, and because I didn't know where else I might be able to get to, I went to the pond. Every hillside was severed by encampments of very low cloud. All was gray and gray-green. Otters had beaten me to it. Fresh droppings were all around the pond, including one on the top of a rock a yard out into the pond. Glistening trails marked their passage through the reeds and the shoreline grasses, and one had squirmed under a fence and left a tuft of fur there. A substantial fish kept surfacing, my first visual confirmation of the presence of fish in the ponds, although the presence of otters, the occasional heron, reports of equally occasional prospecting ospreys all suggested it. Dragonflies were mating on the wet wind, going up together on a cruising spiral. I picked up a blue damselfly from the grass, thinking it was dead, but it stirred lethargically. It sat easily on my finger, the first dry "ground" under its feet for three days. Rain had drenched its wings. I put my hand back on the ground and it stepped off and clung to a rain-bowed grass stalk. I wished us both some sun on our backs.

I rubbed my hand in wet bog myrtle, a deluge of scent. Walking through it is a kind of intoxication. Hours later, the scent would startle me, rising from my clothes as I stood in front of a pub fire.

The pond wears many colours, every shade of blue and gray you could ever imagine, and yellow and gold and red at sunset and sunrise, and almost white and almost black, and on the day of the lull between rains it was treacle-brown for the deluge had heaped the hillside with broken peat and the bloated burn had tumbled tons of the stuff into the pond. The wind creased the surface with tiny ridges of white, so that the impression was of a restless teeming shoal of light-creatures. These seemed to spark off the martins and the swallows and the swifts, for they suddenly descended on the place in loud hunting parties, plundering the surface and the airspace. It was as if a huge mobile had been hung from a low cloud and the wind toyed with it. Pipits fizzed down into treetops where finches crowded and warblers sang. A wren brushed passed my knee where I sat as if I was landscape, and I threw it a passing smile. The pond was galvanised with celebration as the rains relented. The very air felt it too, and it felt clean and clear and fresh and new-born and I drank it in by the gallon, each mouthful cooled by its passage, while every noseful was seduced by...(At this point my notebook records: "interruption to writing caused by d-l-l [a daddy-long-legs] walking down my arm, over my hand and across the open page of the notebook, not pausing to read")...the scent of bog myrtle. The next visitor to my arm was a wee striped hornet. I watched it for a few seconds as it followed the spoor of the d-l-l, but as it neared my uncovered hand I urged it gently on its way. Some of these guys sting and I don't know which ones.

But these are the moments when nature and I come close, when the ancient brotherhood of all creatures and all air and land and water is accessible. There were no long views, the ponds and the rocks were the centre of the world and the cloud-blurred forests of the hillsides were its extremities. My stillness was treated as landscape by everything that moved there. The oak beyond the big pond was the biggest living thing in the world.

I reeled my gaze into the fringes of the pond just below my feet, and noticed an otter dropping I had missed when I arrived, and nearby, a squirming-otter-sized tunnel through the tussocks to the water's edge. I tracked it back and read its movements down a bank, under a fence, across a bog, down a second, smaller bank, round the base of the biggest rock on the shore, and down through the short tunnel into the water. The direction of the flattened grass was the direction of the travelling otter.

"So what then?" I asked myself.

I walked the bank of the pond to the outflow that leads down to the smaller pond and found the otter's exit tunnel. There the flattened grasses lay downhill, marking the otter's passage to its usual entrance into the small pool. The far side of the small pool is where there are almost always fresh spraints so I walked straight there and found the track of the otter resumed. At the reed bed, it all changed, for the grasses showed clearly the otter break into a bounding gallop, the story of its departure held in place and made legible for once by the weight of water that kept the flattened grass flat and the bent grass bent. Beyond the reed bed lay the old railway embankment, and the old curling pond beyond that, a boggy sanctuary thickened with the roots of broken trees for retreating otters to lie up.

The rain came again, not heavy this time but fine and thick, for the sky was still so full of it that it seeped through the seams even when it would not fall heavily. The wetness became too much for the pen on the notebook pages (I was trying to write as I walked), but walking up onto the bed of the old railway – no trains these last 40 years – I found a sharpened pencil, a gift from the rain gods, picked it up with a kind startled gratitude, reopened the book and resumed writing.

I walked in a long arc back to the pond where the new rain hissed two distinct notes – one soft on the open water, the other thinner and higher pitched on the dense pond grasses that had thickened all summer and now in August lay claim to half the surface area of the small pond, and perhaps a quarter of the big one.

"Swans," I told Nature aloud as if she was listening, "to clean the place up a bit you need a pair of swans."

I sat again and the cloud lowered and the sky darkened, and the rain drizzled on and on, and the pond grew gray. I sat on, a long still hour, quite without memorable thought, just being a brother to nature, a ghost to the memory of Henry David Thoreau.

12
The Very Edge of Magic

In the Eagle Glen

And there, becalmed at your feet,
three feathers downcast
and done-with as finished leaves.

Yet these lived
at the cutting edge of flight –
        mile-high spiral
        thousand-feet free-fall
        to a controlled stall.

And there, becalmed in your hand,
the downcast feathers,
but a flick of your wrist stirs

eagle fingertips
and sets the mountain air a-tremble.

Fir-eun, n.m., the eagle, the real bird. – *MacAlpine's Gaelic-English Dictionary, 1832.*

The name of this place is Creag na h-Iolaire, Eagle Crag. Wherever Gaels and eagles have coexisted, there are rocks, crags, buttresses and outcrops with the same name. It is a name that denotes a rock that looks like an eagle, or a rock where eagles are accustomed to perch, or a rock where eagles nest or once nested. The frequency with which it occurs confirms that the golden eagle occupied a significant place in the hearts and minds – and the eyes – of the landscape-namers. Some of the rocks have fallen into disuse since the landscape was named, since the map-makers came inquiring after the name of such-and-such a place, a river, a hill, a rock. But the eagles were there long before the landscape-namers betrayed them to the map-makers, long before the landscape-namers came, and even in those far-back days they used the same rocks for the same purpose. When people came into places like this, they responded to both the golden eagle and the white-tailed sea eagle with two distinct emotions – respect and admiration for their hunting skills and their sheer presence, hatred for the toll the eagles took on their stock and those wild species they could use themselves. But when, by the mid-19th century the Victorians ushered in big sheep farms and sporting estates biased towards big herds of red deer and big coveys of red grouse, they had also begun to take a ferocious toll of birds of prey. This philosophy proved so durable that a 21st century keeper can still utter expressions like "too many hen harriers" and believe he is justified.

The sea eagle vanished from Highland Scotland for a hundred years.

The golden eagle became rarer and rarer, yet it won a reprieve of a kind apparently simply because of something in the character of the bird. Some sneaking sentimental sense lurked here and there even in the breasts of Victorian gamekeepers to whom the sight of a hooked beak was routinely regarded as an excuse to impose the death penalty. And even as the Victorians invented the mania for collecting eggs and stuffed birds, the golden eagle somehow acquired enough friends in high places to thwart the spectre of extinction. As Fraser Darling put it in *The Highlands and Islands of Scotland*:

"Its survival when all the others went under has made it an emblem of the serenity and grandeur of the Highlands."

The serenity and grandeur of the Eagle Glen on a mid-August afternoon in the summer of 2005 was as impressive as it was rare. The upper part of the glen has softened in the time I have known it, a consequence of reversing a decision to plant conifers in the late 1970s to make life easier for the eagles in the late 1990s. The birches in particular have prospered, and in the absence of serious grazing pressure the grasses were a green, waist-high sheen that beautiful summer. It was tempting to look across that rich and fragrant growth and compare it to the qualities you might associate more readily with an alpine meadow in Switzerland. But in reality, it was simply the Highland Edge reverting to type. This *is* a Scottish landscape tradition. We have just lost sight of it for too long. I inhaled its faint aroma of honey that rose on the warm air from dense clumps of tiny white flowers in wet flushes. The heather was poised to burst into bloom and already drowsy with bees. The burn lazed through its easy loops, a shallow, idling water compared to its throaty winter norm. I sat stripped to the waist above the burn, enjoying the warmth and the easy breeze. Before the summer began, it had rained here for more or less a year. But the summer was dry and warm and liberating. Alone in the Eagle Glen on such a day, knowing that at any moment one or both of the adult eagles would rise above the skyline and puts their silhouettes on the day, I felt an almost absurd surge of optimism for this place, for eagles, for wildness. All it takes is time, and the liberation of human imagination from old constraints, the way a benevolent summer liberates a cold-and-rain-drenched spirit and puts the scent of honey on the air.

Sunshine and shadow defined the architecture of Creag na h-Iolaire's skyline bulge. The deepest shadow accommodated the eyrie ledge, for it is both overhung and walled in against the south. There are good times of day to watch the ledge, but this wasn't one of them. It didn't matter. I hadn't come to watch the eyrie. Besides, it was August, and in a good year a newly fledged eaglet would be learning the fluency of wings by now. But it was not a good year for the eagles. For the fourth successive year, and for a variety of reasons, no young bird had flown from the nest. This year a new female laid two eggs but abandoned them in the midst of a relentlessly cold and wet spring, a combination of circumstances that almost guaranteed failure.

This place, this particular Creag na h-Iolaire, is a nest site that has been in continuous occupation for thirty years to my certain knowledge, but my certain knowledge is a blink of little or no account in its eagle history. They could just as easily have been here for three hundred years or three thousand.

The fact that the name itself is old and the birds are still here and use it in preference to two or three nearby alternatives suggests to me that it is a good site in eagle eyes, and one with a very long history of eagle occupation. This despite the fact that in the time I have known the place, the eagles have had more than their share of misfortunes. A succession of failed nesting seasons does not deter them.

By eagle standards then, I am a newcomer to the Eagle Glen with its Eagle Crag, my thirty years acquaintance with the place roughly equivalent to the lifespan of a single golden eagle if it is given the opportunity to die quietly of old age. But it is here that I have acquired much of my first-hand knowledge of the bird, for the territory attached to Creag na h-Iolaire is wholly within my own notional territory, deep in the high hills of the Highland Edge. And whenever I have learned with others in other landscapes – notably with nature writer and golden eagle specialist Mike Tomkies in the West Highlands – it is to this glen, this Creag na h-Iolaire, that I have returned and tested the new knowledge.

I first met the golden eagles of the Eagle Glen long before there was any prospect that I might write about nature for a living, walking these hills on days off from newspaper journalism in Glasgow or Edinburgh or Stirling, learning slowly to read the hills by going alone; learning too to watch with an intensity that is simply not available to the mountaineer with a summit as the be-all-and-end-all of his day. One sighting of a golden eagle crossing a mountain shoulder on wings half-folded and rigid, the landscape patched with old snow, low winter sun pinpointing notes of fire on the nape of the bird, following it in the glasses as it covered a mile of the glen's airspace in less than a minute…all that became summit enough for me once I had acquired the habit of going into the mountains to try and fathom some of their secrets.

Up here, the golden eagle is unchallenged. It has no competitor. The only adversities are the weather, especially in its early winter-into-spring nesting season, and the incomprehensible activities of egg thieves who robbed this eyrie in eight successive years, apparently because one particular female laid pure white eggs, and such a clutch was both immensely collectable and immensely valuable to the thief.

But at that time I was still finding my feet here. I had Seton Gordon's shining example of nature writing, *Days With the Golden Eagle,* for a bible. It begins:

"The golden eagle is supreme amongst British birds in the magnificent power and grace of its flight. It is undisputed chief of the aerial highways.

"The soaring flight of the eagle above the high tops is a thing of great beauty. It seems – and probably is indeed true – that the bird could remain in the air hour after hour without the least weariness."

He watched an eagle soar in a clear sky until it was so high that it simply became invisible to the naked eye. He found a way to calculate how high that might have to be. He estimated 8000 feet. He wrote with a captivating blend of painstaking observation and vivid, picture-making that every now and then touched on fantasy:

"...Always upon that narrow aerial ridge the eagle stands: in winter when all the hills sleep below their soft covering of snow; in summer when at noon the deer in the grassy corrie below drowse in the warm stillness. Here from her lofty perch, the eagle can see hill, loch, and corrie, and westward, peak upon peak merging, in the far dream-like distance, in the last mountain outposts that guard the approaches to Tir nan Og. Does her mind ponder upon those vast distances, great even for her with her unrivalled flight? Or does she watch rather for the swift-wheeling ptarmigan that drift in a white-winged clan across the loch below; or the mountain hares that play upon the hill-face where the heather is not too long; or perhaps the golden plover that speed past on arrow-swift flight calling for the things that have gone away upon the breeze of the hills and will return no more..."

Seton Gordon in that mood is irresistible. More or less coincidentally, I discovered a beautiful book called *Argyll, The Enduring Heartland* by Marion Campbell of Kilberry who would become a dear friend. She too lingered fondly over the golden plover in the high hills in her poem *Levavi Oculos*:

And there, the very edge
of magic, whistling liquidly,
the golden, golden plover wheel and go.

Writing like that had begun to turn my head, and I began to look on golden eagle country in a new light. I began to make regular visits to the same hillsides, intent on encountering eagles there, began to watch where they travelled and how they travelled, began to write down what I saw and what I felt when I saw it. And in the process I formed a personal philosophy for watching wildlife that has stayed with me for thirty years. It is this: that I am as unhappy and confined in a nest-side hide, as I am in the gloom of a badger sett, and that the riches of watching wildlife are mined by becoming a part of the creature's landscape, as freewheeling and out there as the creature itself. Sometimes nature writing relies too much on hides around the nest, around the sett, around the den. It often feels to me like a kind of booby trap that the wild creature walks into, and the results are often dull. Seton Gordon managed to avoid that, made it seem vivid and alive simply by the way he wrote. Eighty years after his *Days with the Golden Eagle* it has still not been better done, and his remarkable nest-side study of the eaglets he called Cain and Abel is counterbalanced by eagle encounters in a variety of Highland and islands landscapes. Mike Tomkies also side-stepped the dangers by articulating both the sheer physical slog of a life devoted to the cause of the golden eagle and an unquenchable love of the bird.

Without ever acquiring the kind of intimacy these two great contributors to eagle literature knew, my own preference is to share wildness with the eagle, not its domesticity, to scan its wide horizons, not a hole in the side of hide in which I pretend not to be there. I *am* there, I want the eagle's passing acknowledgement. I want to be working the same wind the eagle works, I want the same sun on my back, the same cloud shrouds around me, the drenching misery of the same rain when I breast a rise and see the male eagle there on his own Creag na h-Iolaire (not the eyrie buttress but a solitary perching outpost of the territory), looking like a drab rag doll. The golden eagle in particular stirs that kind of longing in me. But exactly the same longing insists that ten years after the event I still have such a profound recollection of an Alaskan bear on his shore, looking for his Brother.

But perhaps the eagles of the Eagle Glen taught me it first. After the eyrie was robbed of its eggs on those eight successive occasions, Don MacCaskill and his wife Bridget decided to organise a watch on the eyrie to try and thwart the egg thieves. I was an enthusiastic volunteer. We watched from the path through the glen far below Creag na h-Iolaire and from up on the watershed. But in later years a wooden hide was flown in and sited a quarter of a mile from the eyrie, and I was less enthusiastic about putting in a shift in the hide. Although it was wind-and-watertight and afforded close-up telescope views of the eyrie and therefore of arriving and departing eagles, I was never comfortable there and resented its restricted view. Up on the watershed I felt like I was *out there*, in eagle country. In the hide, I felt like I was *in there*, in a hut.

So, in the summer of 2005, with a lot of eagle watching behind me, I walked up into the glen on a warm mid-August afternoon. I sat stripped to the waist by the burn and I hadn't come to watch the eyrie and it hadn't been a good year for the eagles. There were no newly fledged young birds acquiring the fluency of wings. It was my third visit in two weeks and on both previous occasions I had seen eagles. I had come back simply because every now and then it seems to matter to me to keep the company of eagles. I feel, in much the same way as Seton Gordon pondered the ridge-perched female's state of mind, that I come closer to the true nature of wildness just by being close to the *fir-eun*, the real bird. It is as if something of that unchallengeable presence of golden eagles transmits itself to a receptive human mind. I am not above asking aloud for a sense of what Fraser Darling called "the emblem of the serenity and grandeur of the Highlands" whenever a golden eagle crosses my path. There is no way of knowing what passes the other way, but it has always seemed to me that if you approach the company of the wildest creatures in your landscape in a respectful way, and if you articulate that in the only way you know how, there is at least the possibility that the creature in question recognises the benevolence of your presence. In the same way, Scott Shelton talked aloud to his Brother Bear.

A week ago, sitting on a high shoulder of the next glen, the male eagle of Creag na h-Iolaire crossed the same shoulder having emerged from his home glen. He saw me before I saw him, and angled away to the north-east when perhaps he wanted east. He crossed the same shoulder where I sat about thirty yards away. His tail stirred, a downward flex of its nearest edge, and he surged forward. His wings were still, and half closed, the primary feathers drawn back behind his tail. Ahead of him was a mile-wide bowl of space contained by the hill wall that dipped towards the wildering north. I lost him eventually against that wall, but by then he had covered the mile of space and perched. And then he had flown again and he was too far off to follow, and I caught myself hoping that whatever had urged his flight, it was not a tight little flock of golden plover on the move towards the very edge of magic.

Two weeks ago, both golden eagles had been in the sky above the ridge, hunting slowly, that ultra-slowness of flight that sometimes borders on mind-numbing stillness, a magic edge of flight. For perhaps ten minutes they appeared and disappeared and reappeared along the ridge, the blackest of silhouettes. Seton Gordon wrote that the Gael called the golden eagle iolairedhubh, the black eagle, and the white-tailed sea eagle iolaire-bhuidhe, yellow eagle. Why yellow I have no idea, but black, yes, that is the silhouette bird of the skyline, the black ghost that Mike Tomkies called "nature's dark angel of death".

And now I sat warm and still, scanning skies and middle distances, quartering the airspaces myself, and over four hours of watching and wandering, no eagle showed. I walked out back down the burn, but paused again by a birch-shaded pool, camera in hand, to try and make something of the burn's bright and spirited descent above the pool, then the wider, deeper darkness of the pool itself, all of it delineated and overhung by the birch trees. But I couldn't persuade the camera to see what I was seeing myself and the light was too intense. So I turned my half-hearted attention to the abstractions formed by the birch reflections in the pool. A patch of blue sky showed there too, and a vivid red rock on the bed of the pool was scarred with short diagonals that seemed to move. Then one of them rose and nibbled the surface. It was a brown trout.

Fish rarely trouble my writing life. I don't spend much time thinking about them. I've never gone fishing, and mostly when I do see wild fish they are dying in the clutches of an airborne osprey. There were perhaps a dozen fish in the pool, the smallest four or five inches, the largest, with a white spot on the top of its head, fourteen or fifteen. The water was warm, warm that is, for a Highland burn. The fish swam slowly and mostly deep in the pool, which was perhaps five feet deep in the middle, but occasionally they strayed out to the shallows. My presence seemed not to trouble them at all. I had just crossed the burn at the lower edge of the pool, I had put down my pack, moved around on the bank trying to compose a photograph, stepped into the shallowest edge of the pool twice, but when I finally saw the fish they still swam in leisurely circles. And suddenly, after the long hours of watching skylines and distance, the glen had shrunk to a small and shaded patch of water and my eyes were looking for fish a few feet below me rather than eagles half-a-mile above. It was a diversion, a mildly interesting attempt to make a photograph out of the play of light on the birch reflections with a hint of the movement of fish whenever they approached the surface. I crouched to try and bring the blue patch of sky more centrally into the composition, wondering idly how it would look if a fish showed against it. I took one fish-less photograph anyway, and decided to wait for a few minutes in case a fish did cross or even pause on the blue.

Then I saw movement nearby and just because I was beginning to enjoy myself – and for no other reason – I held the shutter button down to activate the motor drive. Four or five shots fired as the unmistakable shape of a golden eagle crossed that patch of blue.

I straightened up, and scanned the sky, but the birches crowded close. I ran back up the hill a few yards and scanned the sky again and saw nothing. I stood disbelieving, running the moment back through my mind like a film. There were fish. Then there was an eagle. It crossed the pool's patch of blue as I fired off a handful of photographs. On one of them, it would be there.

It wasn't. The prints showed nothing, other than that a shaded pool on a trout burn on a sunny August afternoon is a good place to waste half a roll of film. My mind still wonders about it though. The very edge of magic, or *fir-eun*, the real bird?

Part Two
The Long Way Back

13
Towards a Native Forest

THE MOST UNCOMPROMISING crossing of the Highland Edge from south to north is by the Pass of Leny. The transition from Lowland to Highland is instant and permanent. The way the sky vanishes and the mountains crowd down on a ribbon of road and a turbulent river made dark by rocks and trees has startled travellers forever. And when the road was but a track and the mountains accommodated wild forests and wolves the sense of entering another land was utter and travellers were as fearful as they were startled. The civilising of the road, the taming of the forests and the extinction of the wolves has not quite nullified the experience. Not even the unsubtle promptings of the tourist trade and the invention of the Loch Lomond and the Trossachs National Park have extinguished the primitive thrill that many a first-time traveller encounters here. And before the landscape eases you into the brighter air of Loch Lubnaig, there is the showpiece jaw dropper of the Falls of Leny. Park at the Forestry Commission's car park, cross the road, walk a few hundred yards and you stand above the falls, where, on the right kind of day, the river thumps between rock walls as boisterous as the sea at Ardnamurchan Point.

For many visitors, this is where they draw their very first Highland breath. Everyone in the tourist business knows the importance of first impressions and the profile of the place has risen because it is now comfortably inside the national park, or, as things stand, uncomfortably inside. As I write this, the welcome signs are covered and travellers have been unable to use the car park for years. The scene around the car park is something between an industrial wasteland and a battleground strewn with the piled torsos and severed limbs of dead trees, for these are the traditional hallmarks of the Scottish forestry industry at work. The environmental importance of the site rarely impinges on the collective conscience of the Scottish forestry industry at work. The haulage roads are crude, the ground is wrecked by the passage of machinery and vehicles, the clear-felling is blatant both from the road and from the hillsides, and all is brightly daubed with warning signs and don't-walk-this-way tape. Where there was a memorable breathing space there is now spectacular ugliness. The overall impression is one of minimal consideration for the landscape setting. The traveller who pauses here now is both startled and baffled. The native, who has seen it all before, is just angry. The scene, or something like it, is replicated throughout Scotland, and in the public imagination, the names of the Forestry Commission and the Sitka spruce are knee-deep in a mire of the commission's own making, knee-deep and sinking.

At the same time, the conservation lobby articulates the case for a renewed native forest. The most quoted statistic – and arguably the most meaningless – is that we have lost ninety-nine per cent of native tree cover, that the surviving one per cent is fragmented into many and widely-scattered fractions of a per cent. Organisations like the Woodland Trust toil heroically over the hill in Glen Finglas to add new fractions of a per cent to the sum total of the fragments with new planting of native species on hillsides where ancient seedbeds still cling on, heroic survivors of centuries of overgrazing by sheep and deer.

The argument, to which I subscribe wholeheartedly, is that if we increase the tree cover we also increase and diversify habitats, and therefore species. A native forest, once it is established, regenerates itself and expands to fill the land available to it. It matches tree species with soil conditions and decides the densities of tree growth. Birds, beasts, flowers and insects follow. But progress is slow, and thus far the scale and impact are small. What we do have is hundreds of square miles of spruce-dominated forest that are routinely clear-felled and replanted with an industrial mentality whose effect on the ground is often loathsome. Yet a simple change of philosophy, a rethink of the national forestry strategy – and especially the scale of the strategy – could quickly transform the landscape while the restoration of native woodland proceeds at its own essentially slow pace.

It can begin right now. We can make a simple plan with what's left of today and go out and start to transform the fortunes of nature in Scotland tomorrow, and the unlikely crux of it all is the Sitka spruce. It is, of course, the species of tree many of us profess to hate more than any other, the *only* tree species many of us profess to hate at all. Almost all of us like trees and many of us love them. Almost all of us do not like the Sitka spruce and many of us profess to hate it. But the fault does not lie with the Sitka spruce. Let it grow *naturally* and it can live for 150 years, and be both beautiful and tall. One on an estate near Perth reached 164 feet. The fault lies in what our forestry industry has done with the Sitka spruce since the commission recognised how straight and tall and quickly it grows. The fault lies in the way it was – still is – planted, conscripted into monocoloured, monocultured, lifeless slabs all across our dearest landscapes. More insults have been heaped on the Sitka spruce than almost any other living thing the length and breadth of the land. Only the midge is more reviled.

Nevertheless, it is to the Sitka spruce that those of us thirled to the cause of our lost forests should turn. At the very least it can be a short-term catalyst for the long-term vision that is a restored and renewed national forest on a national scale. The notion first occurred to me in Alaska, which is where the town of Sitka is, and where the Sitka spruce grows wild, thousands of square miles of it, from the Pacific coast to a natural treeline perhaps 5000 feet higher. That native forest accommodates grizzly and black bear, wolf, moose, flying squirrels, owls, woodpeckers, songbirds, berries, butterflies and flowers beyond our wildest dreams, or at least beyond our wildest memories. Salmon throng the forest's freshwater creeks and bald eagles queue up in their thousands to partake of the feast. And where the spruces wade out into the Pacific backwaters of the Inside Passage, whales, porpoises, seals, sea otters and sea lions are forever cropping up.

But like any *natural* forest species, the Sitka spruce does not grow in isolation in Alaska. In the south-east of Alaska, the temperate rainforest is largely shared with hemlock. Further north, and between the coast and the tundra, the spruce has aspen, fir, pine and poplar for company. It also does not grow in straight lines nor in a uniform density, nor in forests with hard edges, but allows itself space to breathe and grow broad as well as tall; permits clearings, thinnings, thickets, copses, meadows, an understorey, old age and infancy. After three weeks travelling through such terrain the thought occurred to me: this is akin to what Scotland looked like a thousand years ago. Then I began to wonder if what I was looking at could represent a fast track back to a time when Scotland also sustained big forests, big mammals, and the forest reached down and waded out into the terrain of whales.

Ah, but the Sitka spruce is not native to Scotland. It is alien, imposter, fraud, foreigner. That cry is on the lips of every Sitka-hater I ever met, and routinely used among those who hanker after a restored natural forest in Scotland. It is an argument I have little time for. We have, after all, already imported other countries' capercaillie, sea eagles, red kites, red squirrels, and sooner or later we will do the same with beaver, lynx, boar, bear, wolf, the lost tribes of our lost wilderness. We have also imported people from all over the world, and exported them all over the world too. Generations later they turn up in, say, the Jamaican Olympic team: they do not look like us, but they are Campbells, Fergusons, McDonalds, McFarlanes…

And then, there are the Crumleys. My own family, or at least that component of it that conferred my surname on me (there are also Andersons, Barries and Illingworths stirred into the mix), arrived from Ireland around 160 year ago. John and Catherine Crumley were part of a great exodus, and fate and famine washed them up in Lochee, which is the thinking man's corner of Dundee, and for four more generations, Lochee was the tribal lodestone. It says Lochee on my birth certificate. In my own head, I am native, a son of Lochee and Dundee and Scotland, and I never met anyone who felt and behaved and believed himself more Scottish than I do myself. Well, the Sitka spruce has been in Scotland for nine years more than the Crumleys. It was brought back from Alaska in 1831. Perhaps it too has earned the right to believe itself to be Scottish?

So instead of bad-mouthing it as an alien scourge, we could pause to consider whether it might be conscripted to fight nature's cause, and whether for that matter we might treat it better for its own sake and permit it to be beautiful, as it is in Alaska.

If we are willing to do that, the Sitka spruce can be the framework within which a natural national forest of many species can rise and thrive again.

There are two difficulties to overcome. The first is that our forestry industry is a failed one, the pulped end-product unloved and mostly unwanted. The routinely unimaginative re-planting of clear-felled areas at intervals of five or six feet and in straight lines both belittles the landscape and deprives nature of those possibilities we like to call biodiversity. So I suggest we abandon most of the forestry industry.

The second difficulty is that we must transform the *raison d'etre* of the Forestry Commission from industrialist to environmentalist. Its end product would no longer be a pre-determined number of felled trees but the conservation, restoration and expansion of a Scottish national forest where the wellbeing of nature and the diversity of species would be the overriding priority of all its endeavours. Then, the plan is this:

We begin by thinning out plantation forests all over the land and softening their hard edges and straight lines. We do this carefully, treading softly. Wherever possible we use horses and manpower rather than machines, cherry-picking the best trees for nature rather than the pulp mill. Already, the workforce has begun to rise. The Forestry Commission was conceived to create local employment and housing in the countryside. We reclaim that lost philosophy by recruiting locally as much as possible. We train the new workforce in every aspect of silviculture rather than the be-all-and-end-all of cash crop harvesting. We increase the rural population and reinvigorate its social life. A new bond between people and nature has begun, a seed from which countless good things can thrive and blossom.

So we give the trees room to breathe and be beautiful. We make many clearings. Some of these we leave to be just clearings for grass and standing water, for flowers and butterflies and birds and mammals to feed and drink and sun themselves; others we plant with oak, ash, rowan, holly, aspen, larch, Scots pine, juniper, sycamore, maple, gean, willow, hazel, alder, birch (though mostly the birch will look out for themselves, as will the spruces). There will be other species, and not everything we plant needs to have a ten-thousand-year-old Scottish pedigree. Within these many clearings, the young trees prosper because the spruces shelter them while they grow.

We encourage different densities of trees, some areas as open as parkland, some dense for cover, and in time the trees will choose their own densities. We also let the trees find their own treeline. For some species it will probably be somewhere between 2000 and 2500 feet, by which time the trees will be wind-sculpted runts as wide as they are tall. We have one and only one stretch of a working natural treeline in Scotland – at Craig Fiaclach near Loch an Eilean in the Rothiemurchus pine woods – and it is about 300 yards long. We could set a notional target of say, oh, 300 miles. Or 3000?

As the forest begins to mature, we begin to reduce the percentage of Sitka spruce, but the very qualities which make it so commercially attractive also make its presence attractive in a diverse *natural* forest, (for it is as entitled to grow naturally as any other species).

Within, say fifty years, our landscape can be transformed, our tree cover rich and widespread, our flora and fauna more abundant and more diverse. And there is no reason at all, other than the squeamishness of our own species, why wolf and bear and lynx and boar and beaver should not share the forest too.

We will have replaced a sow's ear of a faltering and ugly industry with a silk purse beyond price. We will have created a new rural and stable workforce (these are jobs for all time, for the work of managing the forest to keep it as natural as is humanly possible will never be done), and in the process we will have strengthened the bond between local people and their place on the map. And when travellers encounter the Highlands for the first time at the Pass of Leny and pause at the falls to draw their first Highland breaths, they will be enchanted instead of repelled by what they see. Perhaps they will travel on wondering if and when and where they will catch a first glimpse of a wild wolf standing in the shadow of a towering Sitka spruce.

14
Conservation and Cures

YOU CAN SEE HOW the whole country works from here. So many landscapes of contrast gather so close together. From the mountain summits to the broadest river straths, from the barren heights to the low green lands of plenty, is the matter of a few miles, a few folds in the ground, a few simple steps in the art of landscape making.

At the south extremity of the territory I fondly think of as mine, the land is not twenty metres above sea level. An hour's drive takes you to Glasgow or Edinburgh. At the north extremity the summits are more than a thousand metres. An hour's drive takes you to Glen Coe or the mountains of Atholl. From a low rise in the ground near the south extremity you see the land flatten and grow lush; you see it tilt eastward down the Forth to the far Pentland Hills and the sea. Behind your back is Ben Lomond where the Forth rises, no more than an ooze among rocks and ptarmigan. From the summit of Allermuir, the north-pointing bowsprit of the Pentlands where they lean out above Edinburgh, you see the Isle of May, first and last landfall of the huge firth that begins as that ooze among rocks and ptarmigan, and the May is a haunt of ocean-going puffins and terns. Across the firth stands the improbable Bass Rock where blizzards of yet more ocean-fliers thicken its profile in summer snow: a quarter of a million gannets. And if you turn the other way on the summit of Allermuir you see an arc of blue summits in the far north-west, among which is the conspicuous cone of Ben Lomond.

At the north extremity of my territory, in the middle of the window where my desk contemplates Balquhidder Glen, one mountain prevails. It climbs by a long elegant ridge up out of the Braes of Balquhidder, but in the slow glacial grinding down that shaped the Highland Edge, Stob Binnein's conical elegance was curtailed by one deft slash of the geological axe. So it looks like a volcano, which it is not. Stand on the 1165-metre summit of Stob Binnein and understand how the territory plugs into deeper Highlands to the north and west, for the climb from Balquhidder reaches one of Scotland's pivotal watersheds en route.

These waters that drain south into Loch Doinne and Loch Voil flow into the Balvaig, that flows into Loch Lubnaig, that empties into the Leny, that flows into the Teith that flows into the Forth near Stirling. Those waters that drain north feed the Dochart that flows into Loch Tay that empties into the River Tay that sets sail for Dunkeld and Perth and Dundee, where I was born and grew up looking north across the eastmost arc of the Highland Edge to mountains beyond. On its long march to the sea, the Tay hauls down water from the mightiest catchment area of all Scotland's rivers. It includes even the eastern edge of Rannoch Moor, the fallen-in, melted-down ice cap that once determined the shape of so much of the central and southern Highlands.

And from the summit of Stob Binnein, looking west, your eye is arrested by nothing more shapely than Beinn Lui. And there a tiny lochan near the summit begets the Allt an Rund that joins forces with the Allt Coire Laoigh to become the River Cononish that becomes the Fillan that becomes the Dochart, so it is on Beinn Lui that the Tay rises, no more than an ooze among rocks and ptarmigan. But the western flanks of Beinn Lui drain into the River Lochy and on into the Orchy that fills Loch Awe whose waters course through Argyll to the Atlantic. So here, laid out before me from the summit of the mountain in my window is the full-size map of my life, the life of mountain Scotland, and the visible threads that bind me to it all, that bind my workplace to my birthplace and to both sea coasts. Viewed from here, Scotland is a small country of intense and multi-layered magic.

And yet, not long ago I lived for five years in Glen Dochart (with Stob Binnein's joined-at-the-hip soulmate Ben More in my window), but working with a completely different idea of territory, an idea that felt somehow imposed and therefore temporary. There I often felt restricted by a piece of land no bigger than the territory of a single pair of golden eagles (in fact it was the overlapping territory of two pairs). The book that I wrote there, *Something Out There,* has an introverted feel to me now, and for most of the time I lived there, I felt as if I suffered banishment from the territory that I have subsequently resumed. The temporary territory I staked out never felt plugged into anything. I was rarely at ease there and travelled away from it often and returned reluctantly. When the opportunity arose to reclaim my old territory I accepted it eagerly, and life re-acquired something of its necessary poise. I felt like a lone wolf that had met with hostility in Glen Dochart where the resident pack had fallen on hard times and distrusted strangers, and so I had crossed the old, known watershed where I knew a lone wolf could work well and live at peace.

"But surely," someone said in the bar one night when I was trying to explain some of this, "Glen Dochart is wilder than here. You should have been in your element there." Well the weather is wilder, for sure. The mountains are bolder and the river faster. I saw one or other of the golden eagles from both territories almost daily, and no scanning of the hills failed to unearth red deer. But it has the bareness of deserts, that pared-to-the-bone landscape that is the hallmark of centuries of overgrazing, a land so impoverished its rock ribs show through. Wildness is about giving nature its head, not throttling the life out of it.

Over the years I have watched where nature flourishes and where it struggles, mostly in Scotland (and especially along the Highland Edge, in the Cairngorms and on Skye and Mull) but also occasionally overseas – Alaska, Canada, Iceland, Norway. Studying some of the known details we have lost from the landscape here has helped to make comparisons with these other northern landscapes and reach my conclusions about the worth of restoration, and how we might embark on the long road back to reclaiming most if not all of our own country's natural inheritance. My own era practises what we are pleased to call conservation, and some countries do it better than others. Here on the Highland Edge, we have begun all too tentatively. The red kite has been reintroduced. However misguided the blanket wing-tagging policy may be, the bird is taking its place in the landscape, finding its own niche, settling in. (But in a baffling and potentially disastrous display of uncoordinated policy-making – the curse of conservation everywhere – Stirling Council which administers this patch of the Highland Edge, has permitted a wind farm more or less adjacent to the red kite stronghold around the original release site. It is a question of time before the wind farm starts killing the offspring of reintroduced kites.) The osprey reintroduced itself but we have assisted its recolonisation with artificial nest sites and now it prospers. The Highland Edge is one of its most fertile strongholds. The raised bog of Flanders Moss is unique in Europe and is managed by the Scottish Wildlife Trust. The Woodland Trust is replanting 10,000 acres of mountainside in Glen Finglas with native woodland, crucially building on a few surviving remnants of truly old trees. And the Forestry Commission of all unlikely creatures, has just begun what may prove to be something similar on adjacent land around Loch Katrine. These last two projects alone are sources for some optimism, for they represent a small step away from the prevalent regimes of over-grazing and dull commercial forestry that blight upland Scotland from the English Border to the north coast. But they are small steps, small sources of optimism, and they are the exception rather than the rule. Conservation's potential is constantly reined in by lack of funds, by piecemeal policies, by powerful tourism and landowning lobbies, by politicians' fear of making big decisions and crucially by lack of vision and imagination. Yet my own working territory, with its diversity and core populations of key species, and with a capacity to adopt small populations of reintroduced mammals, could not be better placed to initiate the necessary revolution, to lead by enlightened example. It is not big enough to do the job alone, but if the energies of its fledgling national park could be redirected away from stultifying bureaucracy and an obsession with tourism, to become a showcase for imaginative and expansive conservation, it would have all the visitors it could ever handle, and more important by far, its gospel would spread northwards into heartland Highlands beyond. And there lies a huge piece of country that is the key to everything.

The great shortcoming of conservation in Scotland, in all Britain, and in much of the world, remains its piecemeal, local and uncoordinated nature. It is not a new problem. American Aldo Leopold identified it sixty years ago in his book, *A Sand County Almanac.* Leopold was a professor of game management at Wisoconsin University and much of his written work was scientific. He certainly wrote nothing else remotely like *A Sand County Almanac,* but then neither did anyone else. Its language is original, thoughtful and beautiful. Its messages are wonderfully blunt and uncompromising. The New York Times Book Review wrote of the original edition that it was "full of beauty and vigour and bite". Leopold died helping a neighbour to fight a brush fire a few months after he finished it in 1948. At the time he was an adviser to the United Nations on conservation, but what he really had in mind was rather more than just conservation.

"The practices we now call conservation," he wrote, "are, to a large extent, local alleviations of biotic pain. They are necessary but they must not be confused with cures."

He was only 60 when he died, and from his book it is clear he was enjoying an intellectual golden age. It is difficult to know what progress conservation would have made had he lived for, say, another twenty years, but at the very least its early momentum was stalled by his death. His book has never been out of print, although it is difficult today to discern any impact on the United Nations. But on the Highland Edge, two copies of it occupy this writer's bookshelves. One is a special hardback edition, published to celebrate the centenary of Leopold's birth. The other is a travel weary paperback sent to me by an American friend who found it at a book sale for ten cents and insisted on recycling it. I add to its travel weariness by carting it around on my travels, often in a rucksack pocket. It is immune to spilled coffee, deteriorating forgotten bananas, tomatoes that leak from sandwiches, rain, snow, wind, frost, woodsmoke, midges, clegs. A ladybird once walked across its pages, ignoring the tug of a light breeze that lifted the page where it walked and turned it into a 30 degrees uphill climb. Its pages, like its new owner, are tanned round the edges, the cover is creased at the corners and down the spine and scratched and pounded by God knows how many fingers and thumbs. Two previous owners have left their names on the first page. "Kathy Reid" has signed it in pink ink and "LARSON" has printed his/her name alongside in mundane blue. A mysterious cross in green marker pen is alongside Kathy's name. At the bottom of the page someone has impressed a raven stamp on the paper. What I don't understand is why either Kathy Reid or Larson ever parted company with the book. I won't. You couldn't buy it from me for £100 although I will listen to offers for the centenary hardback that has lived an all-too-sheltered life.

One of the essays in the paperback has been marked here and there in ink, an asterisk here, a tick there, a short diagonal line to indicate the end of a passage that has clearly been purposefully scrutinised. The word "simplify" is written in the margin by a much marked paragraph. The hand-writing could be Kathy Reid's. As one of the book's great attributes is its simplicity, I wondered what on earth it was that must be simplified. The essay concerns Leopold's now famous account of felling an oak tree. What Kathy Reid appeared to want to simplify was this:

"We let the dead veteran season for a year in the sun it could no longer use, and then on a crisp winter's day we laid a newly filed saw to its bastioned base. Fragrant little chips of history spewed from the saw cut, and accumulated on the snow before each kneeling sawyer. [The word "simplify" appears at this point, on its side, so that it lies alongside the following lines.] We sensed that these two piles of sawdust were something more than wood: that they were the integrated transect of a century; that our saw was biting its way, stroke by stroke, decade by decade, into the chronology of a lifetime, written in concentric annual rings of good oak."

Its simplicity is its genius, Kathy.

So an American book travels between owners and ends up dumped in a box at a sale, price ten cents, about six pence on this side of the Atlantic. There it fell under the delighted gaze of the New Hampshire-based painter Sherry Palmer who has owned a copy of the book for many years. Sherry is a friend of mine. She wondered if I had a copy, lavished her ten cents on the possibility that I might not, and a few days later it had crossed the Atlantic and spilled from its envelope onto my desk. My work stopped for two hours, the coffee went cold on my desk, and I leafed through it looking up favourite passages, re-discovering others I had forgotten, the words blazing at me with luminous intensity. I was knee deep in Wisconsin while the world of the Highland Edge went on about its business beyond the window without me for a while.

Here is his conclusion to the book's own little masterpiece, an essay about cranes on a saltmarsh called *Marshland Elegy*:

"The ultimate value in these marshes is wildness, and the crane is wildness incarnate. But all conservation is self-defeating, for to cherish we must see and fondle, and when enough have seen and fondled, there is no wilderness left to cherish."

I think of the red kites with their vivid plastic wing tags and nod vigorous agreement. And his friend John Muir, who I hope needs no introduction to readers of this book, would have nodded too. "To win a hearing for nature from any standpoint other than that of human use, is almost impossible," he wrote, but of course his life's work was dedicated to defeating that very idea.

It doesn't do, however, to glorify such men in soundbites. If we revere them, then we should illuminate the reverence with ideas, with declarations of intent, and with putting them into practice. In the 21st century, if we are serious about conservation on the scale that amounts to cures for all the landscape's ailments, we should be willing not just to think big but also to cherish wildness, to win a hearing for nature for its own sake, not from standpoints of human use. Reintroducing a handful of beavers with radio collars into a fenced test site, prepared to shoot them if things go wrong in the eyes of the local landowner or the local fishing syndicate, is arguably an experiment in conservation – just – but it is light years away from Leopold's idea of cures. Yet that is as far as we have got here. A protracted debate about speed limits for powerboats and jet skis on Loch Lomond centred on the tired and discredited mantra about the need to balance the demands of competing interests. It is a preposterous stand in a national park, for it denies the truism that the landscape's ultimate value is wildness. In a national park of all places, the search for "balance" is the language of the feeble-minded bureaucrats fearful of causing offence and losing their jobs. Management ideals in a national park should centre exclusively on the wellbeing of landscape and wildlife, the cause of wildness, not on the playpark "needs" of people that destroy these things. On the Highland Edge, the policy of wing-tagging every single red kite chick to gather information to fill a report to clutter a shelf in a city office stands conservation on its head. It is a practice designed to satisfy something dull in a few dull people – a standpoint of human use – rather than liberating a species of bird back into the landscape because it deserves to be there. The way to learn about wild creatures, the way to unravel the mysteries that dictate the nature of their lives as a species and as individuals, is to watch them on their terms, not ours. The way to ensure that they survive at healthy, viable population levels is to protect, enhance, restore and extend their habitats. People accustomed to living in the company of wildlife have known these things for centuries, yet biologists routinely decline to accommodate their knowledge and give it credibility. So we reap harvests of documents instead of enlightenment. Politicians take their cue from their advisers and these lean on the politics of the committee room rather than the biology of a salt marsh. And there are no Leopolds, no Muirs, in their ranks.

That is what is wrong. But it is not enough to complain of the wrong-headedness of those charged with the care of the wildness of our land.

The care of the wildness of our land is in the hands of all of us. So how do *we* begin to put it right? How do we begin to unravel the old order, dismantle the moribund structures that stifle progress and inhibit the spread of wildness? How do we prise the land out of the stranglehold the Victorians inflicted on it? How do we render extinct the mentality of the sporting estate that brands as "vermin" every species it finds inconvenient to the pursuit of its selfish corruption of the landscape?

The view from the high hills of Balquhidder that are framed by the window beyond my desk suggests to me that we need to recruit the three best managers of wildness in any northern landscape – the wolf, the beaver, and space.

15
Old Simplicities

I UNDERSTAND VERY WELL the place of the lone wolf in nature's scheme of things. Each time I look in the mirror or consult my own books to jog a particular memory, I detect the spoor of a creature who is not always comfortable with the social niceties and hierarchies of a pack system and so travels mostly alone. Occasionally there will be meetings with kindred spirits or at least sympathetic spirits, occasionally there will be temporary attachments to a pack or at least to its territory, especially when hunting is difficult and an independent resourcefulness stimulates the pack, and the pack's resources provide hospitality for the wanderer. But mostly there is just the wandering.

There is a price. Inner resources can wear thin. Occasionally loneliness sets in like a chill, or a fever. There is the danger of aimlessness. But these are outnumbered by the rewards, particularly if your life's work demands that you seek out the company of nature as a brother. Nature is not an accommodating host to crowds of people. If you climb a mountain in a party of a dozen, you may well share moments of unforgettable camaraderie, accomplish feats of mountaineering, but you will not learn much about the mountain. It will be reluctant to communicate its secrets.

My pack mountaineering phase ended in my thirties. Its end crept up on me, a long, slow dawn that felt like an emergence from a kind of darkness, that lit a different road, a long road back to older simplicities. I began to sense that long mountain days in the company of several other climbers were lacking something. The camaraderie and the summits were no longer my be-all-and-end-all. I would return home increasingly unenthused by the talk of future expeditions to accomplish a difficult section of the list of Munros. In fact, Munro-bagging has always struck me as a kind of madness. I can understand why people collect stamps or antiques or paintings, which are, after all, artefacts of our own civilisation, products of our own ingenuity. But to apply something like the same collectible yardstick to the summits of mountains over a certain height – whether 3000 feet or 8000 metres for that matter – seems to me to be a dismal response to the presence of mountains. I have met mountaineers of distinction who have climbed all over Scotland – some who have climbed all over the world – and who can recall technical details of routes and equipment and climbing techniques and travelling logistics, but nothing at all of birds, flowers, animals, butterflies encountered along the way, nothing of the mountain's secrets. Many of them have a far greater grasp of the peoples of the planet than I have simply because of the nature of their travels. They have seen things and experienced human cultures that I never will. They have forged lifelong bonds with people in faraway places. But a typical mountaineering book is obsessed by the human relationships of the expedition rather than the relationship between the author and the mountain. The mountain is made to feel like a workplace or a conundrum of logistics to be solved rather than a superlative of landscape.

My priorities are different. When I more or less turned my back on pack mountaineering and evolved into a lone wolf, it was because I had begun to understand that what was missing was a deeper relationship with wildness, and that the key was solitude. And when I was much younger, a child on Highland holidays with my family, it was the wildness of mountains, not their height nor even their summits, that first beckoned then spoke to me. Whenever I talk to people about nature writing or even nature watching, I suggest that they should think about going into wild places alone, even if it is only once, even if it is only half-a-mile along the floor of a glen, just far enough to feel untethered, to slip the noose of the pack mentality. Sit somewhere along the way and be still. Listen to the mountain, to its primitive silence, to its winds, its wildlife voices, its river. Drink from the elixir of its wildness. Can you sense how it feels to be a boulder on a hillside or a rock-rooted rowan? Can your mind cast off with the back-flipping raven, the free-fall eagle? What do you feel in your aloneness? Can you respond to these things deep within yourself? You have begun to learn something remarkable – how to keep the company of nature. More importantly, you have begun to re-learn the most precious lost skill your own species ever knew.

But the truly remarkable aspect of my own discovery of the lone wolf's world was that it felt so familiar. I was consumed by some inexplicable sense of having been here before. Thinking about it now, it is still inexplicable. I was neither a solitary nor a lonely child. I could handle my own company when occasion demanded it, but it didn't demand often and I don't remember habitually seeking it out. I have no inclination to delve into reincarnation theories and have never been attracted to the notion nor been presented with persuasive evidence of its possibilities. But there is something in my character that craves old simplicities, that goes looking for elemental things, that wonders how much my species has lost in its relationship with nature's other tribes, and how much of what is lost can be retrieved. When I go alone among mountains, among all wild places, I feel as if I am trying to repair an old and broken connection, like a bridge between landscapes. We broke it when we exterminated the wolf.

Again and again, walking the wolfless mountains of the Highland Edge I feel their absence, or rather I feel the distant and elusive nature of their presence, for no landscape that has sustained wolves ever loses completely the imprint of their reign. As things stand, the golden eagle is the clan chief of Highland predators, and great as it is, it is simply not predator enough. No bird is. Its impact on those two wreckers of the Highlands' natural inheritance, the deer and the sheep, is so small as to be irrelevant. Nor does the golden eagle challenge *us* – your species and mine. There is no discomfiting edge wherever we wander in our wildest landscapes. So although we may revere the eagle and flourish it as a symbol of our landscape, we are never less than comfortable with its presence in our midst. It gives us no pause for thought. In Alaska, Brother Bear at twenty paces gave me pause for thought. I thought:

"My species is not in charge of this situation."

It was the first time in my life that I had been confronted by such a thought, and instead of being fearful I was thrilled by it. That feeling, that awareness, was the single most powerful thing that I have ever learned. And I brought it home with me. I want to be able to feel it here because it belongs here too, because to respond to it in an ancient way is to rediscover the most fundamental of old, lost simplicities, to restore and reclaim our broken and lost inheritance.

The wolf print in the roadside mud gave me pause for thought. I thought:

"He was here, he was, he was *here*, an hour or two ago. What if our paths had crossed? I might have seen him coolly consider my heart-in-mouth presence. We might have locked eyes. What would I have read there?"

I had two conversations in Alaska about *that* kind of presence in the land. One was with Scott Shelton, my guide among the grizzly bears of Kodiak Island. Because the conversation was being recorded for a BBC radio programme, and because I transcribed yards and yards of tape that didn't make it into the finished programme, I can re-read it from time to time to re-live its precious hours, re-learn its precious lessons. These are a few extracts:

Me: It's pretty obvious there's something special going on here between you and the bears. What brought you and bears together in the first place?

Scott: When I first arrived in Alaska in 1974, my work started with Fish and Wildlife [the U.S. government conservation agency] and I spent a lot of time in the field. My first instructor was a Tlingit Indian fellow. Because we were working in bear country I had to get initiated into working in bear country and feeling comfortable. He helped to ease my nerves in my first bear encounters by getting me to really talk to the bears and call them my Brother Bears.

Me: Did you feel awkward about it?

Scott (laughing): Yeah, I felt a little awkward talking to a bear. But he was so calmly speaking to the bear like he was a relative of his... That first season I didn't catch on to the benefits of that and didn't catch on to what he meant, but the more I walked with that individual I saw that it calmed my nerves and calmed the bear as well. The animal wasn't picking up my stress or nervousness...maybe reading me a little calmer, and it helped me to respect the animal a little better by speaking. He didn't have a gun with him any time he was with bears, felt he didn't need one. He felt he could pass through a grass meadow where they come to graze in the springtime, and talk to them as we went through. We would just stroll right through and show them respect. I thought 'What the heck, this seems to work pretty good!' and it's been my philosophy ever since.

Me: Has that got a practical application for what you're doing now, taking complete strangers among bears?

Scott: Yeah, it does help. I try to tell people...you're not foolish to talk to animals. I want the animals to feel the humans are somewhat calm. For some people it can be very stressful to be around their first sighting of a Kodiak bear – any type of bear – but when I bring people in here I want them to respect the animal and to learn from the animal. So it takes some time, but when people leave here they have learned a lot more than they could from reading or movies or whatever. It's reality out here, the way the animals interact, carry on their daily lives. We can just observe. It's a very good learning tool to spend time with any type of animal you're interested in.

Me: It's very salutary not to be the dominant creature.

Scott: I agree with that. The point we try to get across with this programme is that the Kodiak brown bear is a very powerful animal and it needs its habitat. It needs its place where it can be a bear, where it can survive without interference. That's why I put across the point that we have to respect the land they live on. We're walking onto their land, into their habitat. That's why we have to set up boundaries for us to use, and we get rewarded. We get rewarded with the bears coming to see us. The bears decide whether to go round, or how close they want to come. It's very rewarding. Now you have nursing bears thirty feet from you. You're able to see very relaxed bears. It's almost like you're invisible. I know they can smell me. I've got a routine I follow and they respect that. I know they know me, know my voice. I'm not positive, not being able to read a bear's mind, but there's a gentleness there too.

At this point in the conversation I mentioned my enthusiasm for the reintroduction of wolves into Scotland. I wondered if he would consider trying to do anything like this with wolves.

Scott: I'd enjoy doing it with wolves if you get a situation with several good packs. In my time working with wolves, I once spent three hours laying on my stomach in a tidal flat to get this one wolf – I figured he was a leader – to come and check me out. I didn't move for three hours. That wolf took about that long, looking at me from every angle – north, south, east and west – before it approached to about 150 feet. So, I think it would take a lot more time to get a wolf to adjust to your presence, but the only way to learn about any animal is just to be there with it.

Me: If I could take you to Scotland, line up some landowners, farmers, politicians and say this guy can convince you that wolves are not a bad thing, how would you go about it?

Scott: That's a good question, because there's a lot bred into us about wolves – and bears, mountain lions...that they're all out there to destroy everything man has. But they are in the business of making a living too. Like us, they survive on whatever they can survive on. As for reintroducing animals back into their native land, I think it's a wonderful thing for people to realise that there are wild creatures out there and the land belongs to the wild animals, and no matter how big or small they are, it's very important for man to learn to coexist in the same environment. I think the wolf is a good manager. I think the wolf can mange deer, for example, better than man can. And I think on the educational side it would be good to sit back and watch how the wolf manages the deer. And these animals deserve a second chance.

Of course, you might expect someone like Scott to feel that way. He was a man utterly attuned to wilderness life, persuasive proof while he lived of just how much of the ancient bond between man and wildest nature is still recoverable. Establishing a relationship with the Kodiak bears that they would accept was central to his existence there.

But Sherry Simpson was a different creature altogether. She was a journalist and a lecturer in journalism at the University of Alaska in the city of Fairbanks where she also lived. She was cool, sophisticated, confident, quiet, articulate, intelligent, blonde and very good looking. She also had a strong campaigning environmental conscience and as journalist and lecturer she practised what she preached. And she had just been sent on a newspaper assignment to attend a seminar on trapping wolves. At the end of it, the trappers offered this friend of wolves the skin of a newly killed wolf. Tricky, huh?

I was shown her newspaper account of the experience and it intrigued me enough to seek her out. I asked if she ever tried in her writing to unravel the nature of the bond between people and landscape.

Sherry: I do, because it's a way that I can figure it out for myself. Landscape is so much a part of everyone's life here, even people who don't like living here and there are many who don't like living here – I mean, that's a bond too, right? It dominates every part of our lives here. It's also what draws people here and they are often surprised that it's different from what they thought. So I'm curious to know what it is we want from landscape; what will it do for us and what does it say about us? The many ways we respond to it in Alaska are fascinating to me. Some people try to deny it. They'll never go to the woods, they never go hunting, they shop at K Mart and they go to Macdonalds, they try to make something here like everywhere else and what I love about it is here will always overcome that in some way.

After about half an hour of conversation I asked her what she had done with the wolfskin. She looked levelly back at me, and her eyes came at me from a long, long way back. Her voice, which had spoken with control and certainty, was suddenly as distant as her eyes, and in the space between us she stumbled over a sudden absence of certainty. When she did speak, it was more slowly than before. She said:

I have it on a rocking chair in my house. I like to touch it and think about it.

Me: What does it make you think?

Sherry: It makes me think…I wish I could see more wolves in the wild. It's such an incredibly difficult thing to do. It makes me question my own values about things. I'm willing to eat things I didn't kill, I'm willing to wear fur I didn't trap myself, so it makes me wonder where I do stand on a lot of these issues. It's a great reminder of how there aren't really borders here but membranes, and you can cross back and forward between them all the time. That's what wilderness is here. I can live in a city and yet I can somehow cross through that membrane and be in wilderness just like that, and that's important to me.

Her voice and her eyes would stay in my head for months. I thought of her often, moving through the rocking chair room in some humdrum domestic trance between the hoover and the coffee pot, say, absently dragging a hand across the back of the chair and startling herself with the electricity of all that had happened to put the wolfskin there, her mind suddenly vivid with wilderness, a placid city room made tumultuous by the tyranny of a dead wolf. I made a symbol of her wolfskin. It became my shorthand for Alaska. Every time I fingered it in my mind I would startle at its hidden energy, I would question it and long to know more about it, long to feel on my face that keen north-westerly bearing scents and senses of that land where there are no borders, only membranes to pass through at will between the rocking chair and the wilderness; where a bear pauses on the shore of a lake to sense my presence then go on in search of his Brother.

So those were the two conversations I brought back in my head from Alaska. Now, eight years later, looking out from the desk through a windowful of mountainsides on the Highland Edge, the power of what I saw and felt during that too-brief time, the power of that *presence* remains utterly undiminished. The *absence* of anything to engender such powerful forces in these mountains beyond the window remains too as a source of profound regret to me, as it must be for nature. I look at Scott's words on the printed page again and I hear them in my head:

"The only way to learn about any animal is just to be there with it."

The only way for us to learn how to live with the wolf again is to spend time with it. The only way to learn what the animal is really like, as opposed to what centuries of bile and outrageous miscalling have left us with, is to live with it in our midst. And to be willing to give the wolf and ourselves time while it beds down and re-establishes its niche in the land, and to be willing to accept that changes will flow from our willingness, out there among that windowful of mountains.

And it will change things. It will change things in the land and it will change things in us.

16
2015 – A Land Odyssey

DUSK IN THE deep blue-green heart of the Black Wood of Rannoch. The howl of a wolf curls up through the dense and miles-wide canopy, hangs there as a coil of smoke stands and uncoils on a windless day. The sound lingers long after the listening ears of the wood and the hill might have expected it to subside. At last it stills and the silence rushes in and seems to deepen. Then, from far in the north, the answer. In the stillness, the conversation travels five miles. From the forest again, the first voice climbs an octave, and just as it seems ready to fall back it rises instead and holds and holds and holds, then dwindles down, but as it dwindles a second, a third, a fourth wolf combine in eerie concourse, and these stimulate the first wolf again.

The five voices mask from the listening ears the second response in the north but the Rannoch pack hears even as it gathers its five-fold momentum. It hears a single voice in the north, a lone wolf that has wandered into the pack's territory from the northmost of Scotland's three wolf territories, in the Cairngorms enclaves of Rothiemurchus and Glen Feshie. The Rannoch pack falls in behind the alpha male, who sets off at a purposeful lope across the eastern fringes of Rannoch Moor to meet the intruder, if necessary to drive it out, if necessary to kill it, or if necessary to allow it to roam freely along the northern fringes of the pack's territory of around a hundred and fifty square miles.

The moor is no longer the treeless barren land we used to know. It is lightly wooded with young trees, pines mostly, with pockets of birch and juniper, willow and aspen and rowan, the fruits of an enlightened regime that arrived with the wolf, and which the wolf has fostered by keeping the red deer herd constantly on the move. Somewhere south of Loch Rannoch the pack will close on the lone wolf and make a collective decision.

The third pack is in the hills west and north of Balquhidder. Curiously enough, although these live closer to human habitation than the other packs, they are the least seen. My feeling is that there could be only three wolves, although I am not sure. If the intruder on the Rannoch pack's territory could find a way past them to the west, by way of Glen Coe and Etive and across the hills to Glen Orchy, Ben Lui, and on into the Crianlarich hills, it might meet with a warmer reception from the Balquhidder pack than it could expect in Rannoch. It could still be the final outcome, depending on the Rannoch pack's decision.

The Balquhidder pack is the one I know best, which is to say that I have seen the three wolves I know a dozen times since the famous reintroduction of 2010, and heard them twice as often and learned to read the signs of their passage, and to study the patterns offered by word-of-mouth testimony from reliable contacts around these hills. I learned some of the basics of wolf tracking from those two Norwegian film-makers I met in 2004 when a BBC radio project took me into their territory. I have been back several times these last ten years to listen to them and learn from them and travel with them, for they truly know wolves.

They revere the mystery of wolves. They study them on the wolves' terms, using ancient tracking skills and a painstakingly acquired knowledge of the pack's habits. Their half-hour TV film took eight years to film. They fought off the scientists who want to radio-collar and ear-tag the pack. Scientists have collared and tagged all the other Norwegian wolves, so that the pack's movements can be tracked by computer. To achieve the means of their knowledge the scientists must first hunt the wolves from a helicopter and tranquillise them with a dart fired from the helicopter. Old wolves in Norway already know to head for deep cover at the sound of any aircraft engine. I think again of Aldo Leopold:

"All conservation is self-defeating, for to cherish we must see and fondle, and when enough have seen and fondled, there is no wilderness left to cherish."

Norway's wolves reintroduced themselves by walking across the Swedish border. And of course, sheep farmers raged against them for a while. An age-old propaganda of hatred that bears no relationship to biological truth was still in evidence, and no encounter between people and wolves could survive a week without its truth being inflated or made grotesque. But in time wolves found their niche and the Norwegian Government permitted four packs and mostly, the people simply don't see them. And the only wild animal that regularly claims human lives in Norway is the moose.

But the moose is prized by hunters and no-one in Norway wants to exterminate the moose. When you mention the hunters who die most hunting seasons, they shrug and say it's a price hunters must pay. And the moose is so good to eat. Norway's bears, meanwhile (they too crossed over from Sweden but away back in the 1940s) have long since been absorbed into the landscape without fuss.

So from the Norwegians I learned the two basic essentials for the successful integration of wolves into a new landscape: the passage of time and sufficient space.

What made their reintroduction possible in Scotland, of course, was the recognition back in 2007 that the country's two national parks were failing in their primary duties of protecting landscape, restoring and enhancing habitats and restoring and enhancing wildlife populations. In fact they were contributing to the deterioration of these by increasing tourism and development pressures and failing to constrain bad landscape management practice. More important than even that belated recognition was that the Scottish Parliament finally acknowledged the hopelessly fragmented scale of what passed for conservation policy. Even if they allocated major resources and reversed existing philosophy in both national parks, they could see that there was simply not enough space to make worthwhile transformations to the nature of wild Scotland.

I had proposed a solution – a bridge between the two national parks, a third national park, the Heartlands National Park from Glen Dochart and Loch Tay in the south to the Monadhliath in the north and from Atholl in the east to Etive in the west, a huge area of wild country that included Rannoch Moor, Glen Coe, Lochaber and Badenoch. Add to that, the existing national park territories of Loch Lomond and the Trossachs and the Cairngorms, and it was clear at once that there was space to restore habitat on a truly significant scale, and to make plans for the restoration of wildness that had not been attempted before.

The cornerstones of the new threefold sanctuary for wildness were these: the widespread restoration of native woodland, especially pine forest; the reintroduction of the wolf in the Heartlands National Park, in expectation that if they prospered there they could safely expand in two directions (it happened within two years); and the reintroduction of the beaver that had been so timorously treated by the now disbanded Scottish Natural Heritage, surged ahead in six different locations.

So the best three land and wildlife managers in any northern wilderness – the wolf, the beaver and space to breathe and grow – were all in place. The transformation astounded even the most hardened critics with its swiftness. But others had an idea of what was coming from the experience of wolf reintroduction in Yellowstone National Park in the northern United States. After just ten years, in 2005, one student of that project had written in the American magazine *Orion*:

"The wolves...have reshaped huge sections of an awkwardly leaning ecosystem. By pruning the wildly excessive elk numbers, and by forcing elk to be elk again, the Yellowstone wolves kept the herds on the move, allowing overgrazed riparian areas to recover. The elk were no longer encamping in any one spot...and the restored riverbanks served as nesting and feeding habitat for songbirds of different hues. Blink and a howl equals yellow...

"Where previously the overcrowded and static elk and deer herds conspired to keep stands of aspen from regenerating...the beautiful groves of aspen, snow-white bark and quivering gold leaves in the fall, are now prospering, flaring back up the mountainside like so many tens of thousands of autumn-lit candles. Entire mountain ranges are ultimately being painted anew – more colour, more vitality, more light – by the arrival of, initially, a mated pair of wolves, an alpha male and female, followed by the next wave of other wolves..."

In Yellowstone, the wolves had only been gone for seventy years, as opposed to perhaps two hundred in Scotland, although their extermination had been every bit as purposeful and ruthless as Scotland's. Yellowstone showed what was possible with a planned wolf reintroduction, if there was sufficient suitable land available. All that Scotland's campaigners had to do was to demonstrate that such land was available here, and that allied to an overhaul of national park thinking, there was almost no limit to the possibilities and resurgence of wildness in Scotland.

And of course many of the beneficiaries have been people. Much as Boat of Garten proclaimed itself The Osprey Village in the wake of the bird's return to Speyside, so Strathyre in Perthshire proclaims itself The Beaver Village today. You will remember Don MacCaskill's dream about the beavers on Lochan Buidhe (see Chapter 2)? It came true. The first truly successful beaver reintroduction was here. And how they have prospered! Their ability to manipulate the landscape – even one like this that floods so regularly – has had huge benefits for other wildlife species, notably the otters on the river and the swans that routinely saw their nests demolished by flooding before the beavers' arrival. And the beaver visitor centre at Strathyre has transformed the village's fortunes, and become the acceptable face of tourism in the national park.

There have been problems of course. Landowners did not like the notion of a state-owned national park, but realistically, it was the only way to make the project work. Sheep casualties were blamed on wolves regardless of the circumstances of their death. Sporting estates protested that only their fossilised Victorian regime could sustain rural employment (and these now look a bit hollow in the light of the job opportunities the huge threefold park offers to local people, and looking after the regular gatherings of visitors who simply want to hear a wolf howl).

It is perhaps a measure of how far things have come in such a short time, however, that the biggest single outcry arose when beavers found their way up the River Balvaig and through Loch Voil in Balquhider Glen to Loch Doinne at the head of the glen, and began to colonise there. It wasn't the presence of the beavers that caused the outcry but rather that one of their number appears to have been killed by a wolf. Only then did lingering vestiges of the old propaganda surface.

But as I have said, these Balquhidder wolves are the most elusive of all, and no matter that much of my life is now spent in pursuit of them, walking the hills and holing up close to their preferred resting places and hunting grounds, the pack has proved to be very sparing with its secrets. And for very good historical reasons it is distrustful of the goodwill that now exists here and there in these hills, and in the huge and open lands to the north. I e-mail my friends in Norway, asking for help. They send me back a shrug, and tell me to go out again and use my eyes better.

And of course the irony of my own situation does not escape me. Twenty years ago, there was no more vociferous opponent of Scottish national parks. And when the first two had been set up and blundered clumsily through their early years, I was among those who loudly protested that all our worst fears had been realised. Their management was spineless and ineffective and wasteful, and their landscapes were poorly served. Championing the cause of wildness simply did not figure in management philosophy.

Slowly, though, I began to see that there could be purpose, if only the management was handed over to people with a proven track record of national and international conservation, and if only the scale of land available to wildness was sufficient to make a difference.

So the Heartlands National Park became the key to everything. Now each of the parks has its own local management board charged with maintaining the individual characteristics of the area, resolving local issues and promoting environmental education, which has the worth of wildness as its first commandment. But there is also an overall management authority that develops and implements policy throughout all three parks, and this body reports directly to the Scottish Executive's Department of Wildlife and Landscape at their new offices in Killin. The Minister and the Department staff constantly visit the three parks to study work in progress at first hand.

The world beats a path to our door to see the transformation we have worked and to find out how it has been possible. And sometimes just to sit on a hillside above the miles-wide blue-green heart of the Black Wood of Rannoch to listen to the howl of a wolf that curls up and hangs above the trees as a column of smoke stands and uncoils on a windless day.

That is how the world looks to me high on a shoulder of Stob Binnein of Balquhidder, one late spring evening of 2015.

Epilogue

That Day Summer Died

That day, summer died
of old age.

Cloud and mountain
piled and whitened.

Wind suddenly cool enough
for swans' wings

out of the north-west
stirred the loch to yawns.

Eagles opened golden doors
for northern swans.

Loch Doinne, Balquhidder
October 2005